Longman
Geography for *CSEC*®

Jeanette Ottley & Marolyn Gentles

T0173220

Hachette UK's policy is to use papers that are natural, renewable and recyclable products and made from wood grown in well-managed forests and other controlled sources. The logging and manufacturing processes are expected to conform to the environmental regulations of the country of origin.

Orders: please contact Hachette UK Distribution, Hely Hutchinson Centre, Milton Road, Didcot, Oxfordshire, OX11 7HH. Telephone: +44 (0)1235 827827. Email education@hachette.co.uk Lines are open from 9 a.m. to 5 p.m., Monday to Friday. You can also order through our website: www.hoddereducation.com

First published by Pearson Education Limited
Published from 2015 by Hodder Education,
An Hachette UK Company
Carmelite House
50 Victoria Embankment
London EC4Y 0DZ
www.hoddereducation.com

2024
IMP 10 9 8 7 6 5

ISBN: 978-1-4058-1663-2

Prepared for publication by Cathy May

Illustrated by Tony Wilkins

Printed and bound by CPI Group (UK) Ltd, Croydon, CR0 4YY

With thanks to Peter-John Gentles for his help in producing the original artworks.

Index prepared by Indexing Specialists (UK) Ltd.

Contents

We are grateful to the following for permission to reproduce photographs:

A1Pix: p.198; **Alamy:** p.1(tr) (John James), p.1(bl) (Worldspec/NASA), p.27 (Mike Kipling), p.65(t) (URF/f1online), p.66 (Douglas Peebles Photography), p.109 (Reinhard Dirscherl), p.111 (Brandon Cole Marine Photography), p.174(bl) (Bill Bachmann), p.207 (Oliver Benn), p.253(tl) (Mark Bacon), p.253 (br) (Andre Jenny); **Caryn Becker Photography, photographersdirect.com:** p.164; **Corbis:** p.1(br) (Stapleton Collection), p.65(bl) (Philippe Giraud/Goodlook Pictures), p.175 (tr) (Morton Beebe), p.205 (Macduff/Everton), p.244(l) (Stephen Frink); **Corbis Sygma:** p.246; Ecoscene: p.244(m) (John Liddiard), p.244(r) (John Liddiard), p.253(bl) (Erik Schaffer); **EMPICS:** p.154(b) (Sukree Sukplang/AP), p.217(b) (Bullit Marquez/AP); **Eye Ubiquitous:** p.158(Bruce Adams), p.175(tl) (David Cumming), p.181(Mike Atkins); **GeoScience Features/Dr Basil Booth:** p.70, p.78; **Robert Harding:** p.100 (Robert Francis), p.189 (Tony Waltham); **Hutchison Library:** p.174(r) (Sarah Errington), p.179 (Tim Beddow), p.217(tl) (Philip Wolmuth); **Lonely Planet Images/Alfredo Maiquez/Getty:** p.197; **MCA Photography, photographersdirect.com:** p. 108(b); **Panos:** p.65(br) (Rob Huibers), p.149(Marc French), p.193(Pietro Cernini), p.217(tr) (Rob Huibers), p.247(r) (Dieter Telemans); **Kevin Philips Photography, photographersdirect.com:** p.148(l); **Photographic Art Jamaica, photographersdirect.com:** p.160, p.173, p.228; **Punchstock:** p. 1(tl); **St Vincent & The Grenadines, Forestry Commission:** p. 257; **David Simson:** p.148(r), p.175(c), p.175(b), p.209; **Still Pictures:** p. 55 (Qinetiq Ltd), p.255 (Mark Edwards); **Topfoto/ImageWorks:** p.182; **Travel Ink:** p.126(r) (Frederick Salgado); **TRIP:** p.150 (D Saunders), p.154(t) (A Tovy); **tropix.co.uk:** p.174(tl) (Martin Fleetwood), p.183 (Veronica Birley), p.236 (Lyn Seldon).

We are especially grateful to Marolyn Gentles for the use of her photographs.

Picture research by Louise Edgeworth.

Every effort has been made to trace the copyright holders and we apologise in advance for any unintentional omissions. We would be pleased to insert the appropriate acknowledgement in any subsequent edition of this publication.

CHAPTER 1

Examination techniques

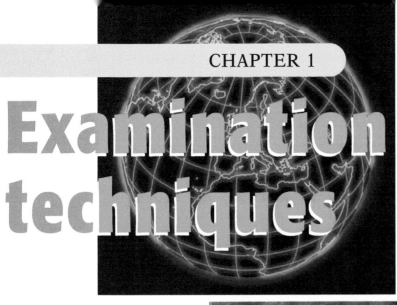

By the end of the chapter students should be able to:

- define **Geography**
- explain the **structure of the CXC Caribbean Secondary Education Certificate (CSEC) Geography examination**
- practise **examination techniques**.

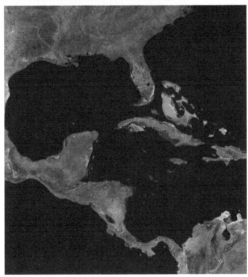
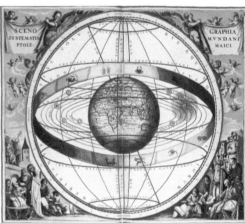

What is Geography?

✳ **Geography** is the study of places on Earth and the interrelationships between natural and human environments.

You are probably preparing for examinations in many subjects. Each subject has its own approach to its content matter. For example, a writer, geologist, engineer and geographer may study a cliff in different ways. So it is important to understand the subject in which you are being examined.

The central question of Geography is: **Where**? We then ask: **Why there**? Many examination questions will ask you to draw maps to show location, or diagrams to describe geographical features. You will often be asked to explain the factors that influence the location of these features.

Students with a Certificate in Geography can go on to work in many career areas. Many geographers are employed in environmental work as consultants, educators, engineers and landscapers. You can study Geography at CXC Caribbean Advanced Proficiency Examination (CAPE) level with many other subjects, including sciences and arts subjects. You can gain a CAPE Associate Degree in Environmental Science with Geography as one of your subjects; or in General Studies with other subjects.

The CSEC Geography syllabus, and this book, are divided into four main sections:
I   Map reading and field study
II   Natural systems
III   Human systems
IV   Human–Environment systems.

Geography uses the **field study method** to investigate the natural, human and human–environment systems. The Field Study Report forms your compulsory School-Based Assessment (SBA). It is awarded 20% of your final marks.

The main tool of the geographer is the **map**. Its importance is underlined in the examination as map reading is a compulsory question and, with other practical skills, accounts for about 30% of the final assessment. The use of maps and field studies should be practised throughout your two-year course of study.

You will be given a grade on your performance in three areas called **profiles**:
Profile (PS)   Practical Skills
Profile (KC)   Knowledge and Comprehension
Profile (UK)   Use of Knowledge.
(Each subject has its own profiles.)

Go to www.cxc.org to find more resource materials.

In this book, separate chapters deal with these important topics. For example, in Unit I (Map reading and field study):
Chapter 1   Examination techniques
Chapter 2   Map reading and other practical skills
Chapter 3   The School-Based Assessment.

In preparing students for CSEC Geography the book embraces the use and awareness of technology so essential for a young person in the twenty-first century. Throughout the book many website addresses are given to encourage further research.

Examination structure

Being successful in examinations at this level requires three things:

- a thorough understanding of the content required by the syllabus
- practice in the skills and specific objectives required by the syllabus
- familiarity with the exam format and the time allocation.

You may use a non-programmable calculator for this exam.

CSEC Geography is assessed by two written papers accounting for 80% of the marks, and an SBA worth 20% of the marks.

Note: Failure to submit the compulsory SBA results in automatic failure of the subject – you will be ungraded (U).

You should consult your teacher very early on in order to choose a good and manageable topic. The SBA field study is not a project as in other subjects. The Geography SBA requires you to go out into the field and observe, measure and record some geographical phenomena. You need to carefully select and prepare your project for study (see Chapter 3).

➤ *Figure 1.1 Field study – escape from the classroom*

You are required to complete the following:

Paper 1   60 multiple choice questions in $1\frac{1}{2}$ hours (30% of the final assessment)

Paper 2   A compulsory map reading question, and one from each of three questions set from each of Natural systems, Human systems and Human–Environment systems – a total of four structured questions in $2\frac{1}{2}$ hours (50% of the final assessment).

Longman Geography for CSEC covers the entire content of the syllabus including detailed regional and extra-regional examples. It follows the 'systems' approach of the syllabus and includes highlighted concept boxes. The specific objectives of the syllabus are placed at the start of each chapter for easy reference.

There are exercises to help ensure mastery of the material throughout the book, and sample questions for each exam paper at the end of each unit.

Answering questions: Paper 1

Paper 1 consists of 60 multiple choice questions to be answered in $1\frac{1}{2}$ hours.

24 questions are based on practical skills (PS), 28 questions mainly test your knowledge of definitions and concepts (KC) and 8 questions test use of knowledge (UK).

Note: No multiple choice questions are set from the parts of the syllabus that include a choice of territory to be studied.

You will be given a grid with letters A, B, C, D, and you should shade *one* letter corresponding to the correct answer. You will need a soft 2B pencil and a clean eraser – these questions are mechanically marked, so your intention to shade one answer must be clear. If you shade more than one, your answer will be given as wrong.

Each set of answers for a multiple choice question includes an obviously wrong answer (detractor), two others which may be partly right or sound right, and the correct answer (like the television game show 'Who wants to be millionaire?'). If you do not see the right answer straight away, you can eliminate the detractor and any other answer that includes wrong parts (like the 50:50 lifeline) and end up with perhaps two answers from which to choose.

You should try to answer each multiple choice question in 1 minute, in order to give yourself 15 minutes to check any questions you may have missed.

Sample multiple choice question – Paper 1

Exercise

All the following landforms are found in the lower course of the river:

(A)   flood plain, delta, river cliffs

(B)   delta, meander, ox-bow lake

(C)   wave-cut platform, cliffs, V-shaped valley

(D)   delta, interlocking spurs, bars

If you do not immediately know the correct answer, then you can eliminate obviously incorrect ones. (C) is the detractor as it includes coastal landforms. (A) and (D) include landforms from the middle and upper course (cliffs and interlocking spurs respectively) and so are only partly correct. The correct answer is (B) – delta, meander and ox-bow lake are all features of the lower course of a river.

You will find sample multiple choice questions at the end of each unit in this book.

Note: Never leave out a multiple choice question – you have a 25% chance of being right whichever answer you choose.

Answering questions: Paper 2

Paper 2 consists of 10 questions from which you must answer *four* in $2\frac{1}{2}$ hours.

Question 1 is on map reading. Map reading is very time-consuming. Some students may wish to do this question last. Do not spend more than 45 minutes on it. It carries 28 marks.

You will be given a map extract, often of a Caribbean territory. (If it is your own territory, be careful not to include any information that is not given on the map. Your answers to these questions must all come *from the map only*.)

The following is a general guide – the actual questions in any one year will be different.

- 8 marks can be given for questions testing your ability to measure distances, give grid references and directions – usually just 1 mark for each answer, so do not spend too much time on them.

- 6 marks can be given for calculating a gradient, drawing a cross-section or a scaled sketch map. (In any one year, not all of these may be tested.)

- 6 marks can be awarded for your use of map evidence to describe/identify landforms, drainage, vegetation, land use, settlement and communications.

- 8 marks can be allocated for explaining relationships between features such as relief, drainage, vegetation, settlement, land use and communications.

It is important, therefore, to distribute your time carefully according to the allocation of marks for the total of 28 marks (see Chapter 2).

For the **other three questions** you choose *one* from each of three questions set on each section:

Section B   Natural systems – you choose *one*.
Section C   Human systems – you choose *one*.
Section D   Human–Environment Systems – you choose *one*.

Each of these questions will be set from the content of the syllabus for that section. You should spend not more than 35 minutes (that is, about 1 mark every 1.3 minutes) on each question.

Each of these questions has parts testing the three profiles of the subject:

- Profile (PS)   Practical Skills – 4 marks

- Profile (KC)   Knowledge and Comprehension – 8 marks

- Profile (UK)   Use of Knowledge – 12 marks.

Each part of the question will have a *command word* – this tells you what you have to do to get the marks.

Command words

- **Practical Skills** (4 marks): you will be asked to do something like '*Draw* a map/diagram and *locate* on it …' or '*Use the given table* to answer … ' (see Chapter 2). All responses must be given on the map or diagram drawn, or answers taken from the given data.

- **Knowledge and Comprehension** (8 marks): you will be asked to *list*, *name*, *state* – these terms require you to answer simply in one or two words. Often 1 mark is awarded per answer. Do not waste time describing or explaining if this is not required. You do not get more marks for writing more words.

  The command *describe* requires a bit more writing. You would need one or two sentences to properly describe the concept or process. Look at the mark allocation: if you are describing two processes for 4 marks, then you need to make at least two statements about each to gain full marks.

- **Use of Knowledge** (12 marks): you will be asked to *explain*, *compare* or *analyse*. This type of command is testing your higher-order understanding of the subject, not mere recall. It is difficult to 'cram' for this part of the question. It requires sustained writing in paragraphs outlining the main points. It should include use of relevant examples. Often, 3 or 4 marks will be allocated for each item, for example, 'Explain *two* factors influencing the distribution of population in a named country. [8 marks]' You will need to give 4 points (sentences) about each factor.

These questions will relate to a named place, so your answer must be specific to that place. For example, all cities function as business centres of some sort and most cities have a central business district, but Kingston has two business areas. Comparisons must be based on specific knowledge.

● **Diagrams, maps** and **sketches** may be used in any questions to improve your answer. Marks will be awarded if they show additional material not already given in the written part.

Note: If a question has the command *with the aid of diagrams*, this means that marks are allocated specifically for the diagrams, which should be well labelled. Even if you write well, you will not gain those marks without diagrams.

Look at the mark allocation for each section of the question – this will be a guide as to how much you should write.

1 You must know the **content**. If the question is on 'hazards' and you have not studied hazards, then do not choose that question. (In Geography you have to be able to analyse the content and refer to specific case studies/examples.)

2 You must be able to do all the **commands** – *draw, describe, explain*.

3 You must be able to answer **all parts** of the question.

➤ *Figure 1.2 Choosing a question*

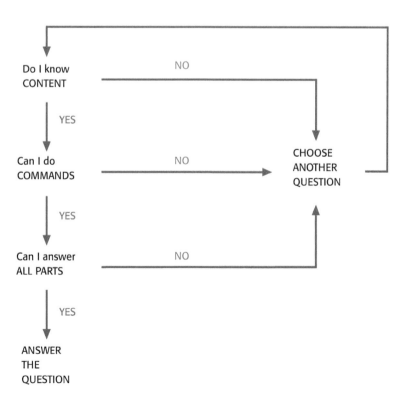

7

Some examples

(a) Draw a well-labelled diagram to show a wave-cut platform. [4 marks]

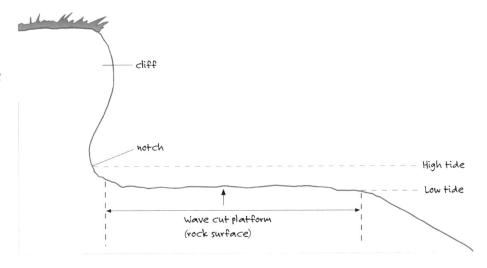

➤ Figure 1.3 Example correct answer

(b) Describe TWO characteristics of fold mountains in the Caribbean.   [4 marks]

➤ Figure 1.4 Example correct answer

I Fold mountains in the Caribbean are characterised by their height. They generally rise to over 1000 m as in the Northern Range of Trinidad or even more than 2000-3000 m in Jamaica and Cuba. They form extensive mountainous areas.

II The Caribbean fold mountains are very steep-sided. The original complex folding formed steep anticlines, but later erosion by rivers and landslides also resulted in numerous steep valleys and spurs. The fold mountains of the Caribbean are relatively young and very steep.

(c) (i) List TWO processes of chemical weathering.   [2 marks]

➤ Figure 1.5 Example correct answer

Carbonation. Solution.

(ii) Describe biotic weathering.   [2 marks]

➤ *Figure 1.6 Example correct answer*

Biotic weathering is the break-up or decomposition of rocks by plants. Plant roots can grow in cracks of rocks and break them up. Humic acid formed from decaying vegetation can also chemically weather the rock.

(d) Explain how I levées and II ox-bow lakes are formed. [8 marks]

➤ *Figure 1.7 Example correct answer*

I   Successive flooding of the river forms levées. The flood water overflows the banks of the river and spreads out over the valley floor. The heaviest sediments are dropped first, closest to the river channel. (The finer sediments are carried further away and spread over the flood plain.) Repeated deposits of larger sediments build up banks higher than the flood plain. These pairs of raised banks are called levées.

II Ox-bow lakes are created when pronounced meanders are separated by a narrow neck of land. Erosion occurs on the outside of the bend and deposition on the inside of the bend. Eventually river erosion cuts through the neck of land leaving the meander apart from the main channel. Deposition further seals off the meander. This forms a crescent-shaped temporary lake called an ox-bow lake.

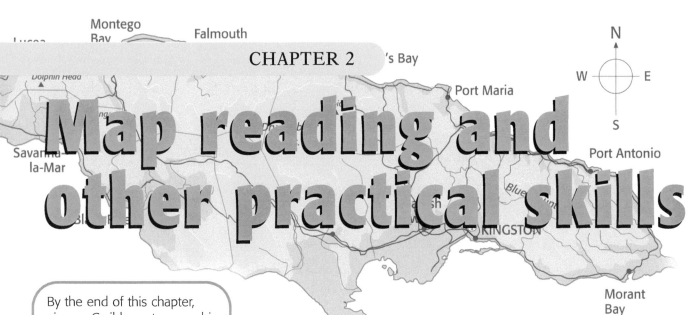

CHAPTER 2

Map reading and other practical skills

By the end of this chapter, given a Caribbean topographic map students should be able to:

- locate places, using **four-figure and six-figure grid references**
- give directions in terms of **compass bearing** and the **16 points of the compass**
- use **scale** to measure distance
- **reduce** and **enlarge** a section of the map
- describe **landforms** through the reading of contours
- draw and interpret **cross-sections** and **sketch sections**
- calculate **gradients** using ratios
- read and interpret **conventional symbols**
- describe **drainage, vegetation, land use, settlement** and **communications**
- explain the **relationships** among **patterns of relief, drainage, vegetation, land use, settlement** and **communications**.

Students should also be able to:

- interpret **geographical data** from a photograph
- locate a place from its **latitude** and **longitude**
- find the **latitude** and **longitude** of a **given place**
- locate **territories** in the Caribbean
- calculate the **time** at a place
- draw **sketch maps** to show relative location and spatial distribution
- draw **diagrams** to illustrate geographical features
- construct **line** and **bar graphs** and **divided circles**
- interpret **tables, dot maps, choropleth** and **isopleth maps, bar graphs, line graphs** and **divided circles**.

Note: This chapter should be used as a reference for all other chapters throughout the course.

Map reading and practical skills are very important to the study of Geography. **Location** and **distribution** are recurring themes in Geography. A geographer must be able to *observe, record, present* and *interpret* data in many different ways. In fact, these skills are considered so important that almost one-third of the total assessment marks of a student's performance is measured in this first profile called 'Practical Skills'. The skills given here should be applied to all areas of the syllabus. They are used in the compulsory School-Based Assessment field study (Chapter 3) and account for 25% of its marks.

This chapter is divided into four sections:

● map reading

● photograph interpretation

● global positioning and sketch diagrams

● statistical diagrams.

Map reading

> ✳ A **map** is a plan drawn to scale, showing the arrangement of features of the Earth's surface from above – a plan view.

Maps are the basic tools of Geography. The drawing of maps is called **cartography**. The oldest known map is Babylonian and dates from 8 000 years ago. This map was inscribed on a clay tablet and showed the positions of houses on terraced hillsides with an erupting volcano in the background.

In more modern times, the use of surveying instruments has resulted in accurate maps of the whole world. In the twenty-first century, orbiting satellites take accurate aerial photographs which can be used by computers to produce up-to-date accurate maps of the world. Other spacecraft have even produced surface maps of some planets. Geographical Information Systems (GIS) are able to show the detailed distribution of the Earth's features.

There are many different types of map, for example:

● Dot/choropleth maps can be used to show distribution of settlement, livestock, crops.

● Topographical maps show the landforms and human activity on the Earth's surface.

We will be using topographical maps in this section. Map reading and interpretation should be practised throughout the year. It should be applied to every topic on the syllabus. You will be required to draw two maps for your SBA (see Chapter 3).

❋ A **topographical** map is one that shows the relief of the land (topography) as well as all the human features located on it.

➤ *Figure 2.1 Topographical map with scale, key and north arrow*

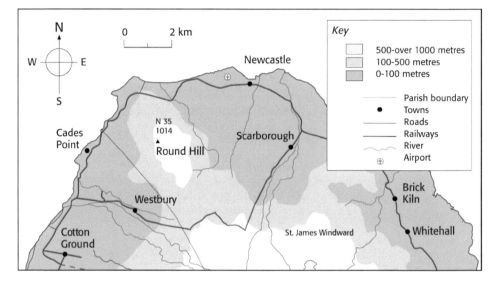

Grid references

A point/area on a map can be located using grid references. Grid lines called eastings and northings are drawn over maps at right-angles to each other. Lines drawn vertically (top to bottom) are eastings with numbers increasing eastwards; lines drawn horizontally (across) are northings with numbers increasing northwards. We can locate any place on a map by stating where it is along the eastings and northings.

➤ *Figure 2.2 Grid references*

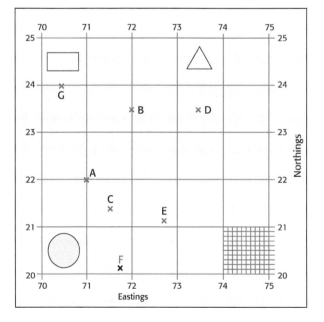

On Figure 2.2:

- Point A is located where easting 71 meets northing 22; we can write its grid reference as 7122. But if a point lies between the lines, we divide the square into tenths and write a third number for each reference to give a six-figure reference – 710220.

- Point B lies on easting 72 and between northings 23 and 24. We estimate the distance between B and line 23 in tenths and state the grid reference as 720235 – that is, a six-figure grid reference.

- Point C lies between eastings 71 and 72 and northings 21 and 22, so we need to estimate where it lies in both directions; the grid reference is 715215.

Note: Eastings are **always** written first and there must be an even number of digits.

Four-figure references are used to show areas. For example, 7424 refers to all the points from 74 up to 75 and all the points from 24 to 25 but not including 25 itself. (This area is shaded on Figure 2.2 – the first set of numbers is for eastings, the second for northings.)

Six-figure references represent a point where the lines intersect.

Exercise

Use Figure 2.2 to answer the following questions.

1 Write the six-figure grid references for the points D to G.

2 Identify the shapes at the following four-figure grid references:

7020   7324   7023

Compass points, directions and compass bearings

The relative position of one object on the map to another, or its direction, may be given using the 16 points of the compass, or compass bearings.

The 16 points of the compass are those given as north/south, east/west and the divisions between them (see Figure 2.3).

Compass bearings are angular readings between two points. (In the field, you would use a compass to give the bearing of the point from your own location.) The method of giving bearings is shown in Figure 2.4.

It is very important to note:

- which point you are taking the direction **from** – this is where you are, looking to the other point you are taking the direction **of** .

- the direction of **north** on the map (conventionally north is to the top of the map, but there should be an arrow identifying north on the map)

- all compass bearings are taken **from north** in a **clockwise** direction.

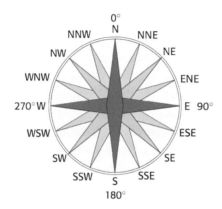

➤ *Figure 2.3 Compass points and compass bearings*

Exercise

You can make a simple compass diagram by copying Figure 2.3 onto tracing or other transparent paper.

As you can see, the 16 compass points do not give the exact direction of all points. If you use a protractor, you can give the more accurate direction as a bearing between any two points.

Steps in giving directions:

1 Join the two points by a pencil line.

2 Note which point you are taking the direction **from**.

3 At that point draw a line parallel to the north arrow on the map.

4 *Either* compass direction: Place your simple compass diagram at that point. Align it with north and read off the direction to the point along the first pencil line. The answer should be given as a compass point, e.g. north east.

   Or bearing: Put your protractor with zero along the north line and measure in a **clockwise direction** to the pencil line. The answer should be given as an angle (three digits), e.g. 034 degrees.

Note: A circular protractor with 360 degrees is more useful than one with 180 degrees.

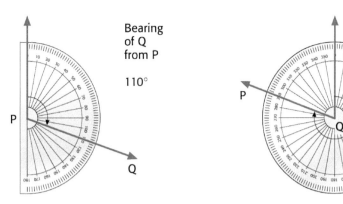

➤ *Figure 2.4 Taking a compass bearing*

Bearing of Q from P

110°

Bearing of P from Q

290°

Exercise

Use Figure 2.2 to answer these questions.

1 What is the direction:

   a of B from A?   b of E from D?   c of G from A?

2 What is the compass bearing:

   a of A from B?   b of D from F?   c of E from B?

Note: If the question asks for 'direction', give the answer as a compass point; if it asks for a bearing, then give the answer as an angle.

Scale

The scale of a map is one of its most important feature. The scale states the relationship between the size of an object on the map and its real size. If the object on the map is half its actual size, then one unit on the map represents two units in real life. An object on the map would be half its actual size.

✷ A **scale** is a statement of the relationship between the size of a feature on a map and its actual size.

The scale may be written **in words**: 1 centimetre on the map represents 1 kilometre in the real world. But this limits its use to those who understand the language.

The scale as a fraction could be used by anyone. The top number of the fraction represents the map and the bottom number the 'real world'. This fraction is called the **representative fraction**. A fraction or ratio of 1:10 000 means that one unit on the map represents 10 000 units in the real world.

Note: Any unit may be used but both sides of the ratio must be in the same unit.

If we use metric units then we can write 1 centimetre on the map represents 10 000 centimetres in reality. We measure the real world in metres and kilometres, so 10 000 centimetres can be converted to 100 m. So a distance of 10 centimetres on the map of scale 1:10 000 will actually be (10 x 10 000) centimetres long, or 1 000 m.

(Some maps use imperial units – inches, feet, yards and miles – to indicate distance and heights. So these are written on the map as *heights in feet*. There are 12 inches in a foot, 3 feet in a yard and 5280 feet in a mile. You should use *one* system of units for any map.)

Many maps have a **line scale**. This is a line drawn accurately to the scale of the map, with the units for the real distance marked off to the right of the zero and fractions marked off to the left of the zero. Any distance on the map can be worked out by simply placing a length of paper or string measuring a distance from the map on the line scale, and reading off the actual distance in the real world.

Measuring distances

The scale of a map can be used to measure distances between points or along roads, or lengths of rivers.

Straight-line distances may be measured by calculating in words or numbers as described above. A piece of paper with the points marked off from the map may be placed along the line scale: the point on the right is placed on the nearest whole number and the fraction is read off to the left of the zero.

Distances that are not straight may be measured by laying a piece of string along the route on the map and marking the start and the end. The string is taken up and straightened out on the line scale and the distance read off the scale as above.

Exercise

1 Write the following scales as a ratio:

   a one centimetre represents one thousand centimetres

   b 3 centimetres represent 3 000 metres

   c one centimetre represents half a kilometre

   d one inch represents 25 000 inches

2 Express the following representative fractions in words (metric units):

   a 1:50 000

   b 1:10 000

   c 1:25 000.

3 Use the line scale in Figure 2.1 to work out the distance represented on a piece of paper with 3.5 centimetres marked off on it.

4 Given a scale of 1:100, measure the distance in metres between points A and B on Figure 2.2.

Exercise

Draw a simple map of your classroom to show the relative positions of the desks and teacher's desk. You may choose symbols to show the blackboard, windows and doors. Remember to include a title, frame, north arrow (look at the direction of the morning sun – that is the east) and scale on your map.

Reducing and enlarging a section of the map

A map may be reduced or enlarged by changing the size of the grid lines and then copying the information in each square.

If we wish to enlarge a section of the map, we may make the grid lines twice as large and then copy the features from each square onto the large grid as required.

➤ *Figure 2.5 Enlarging and reducing maps*

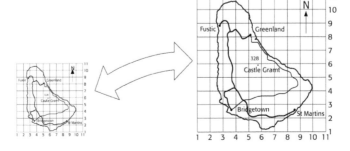

➤ *Figure 2.6 Map of Nevis*

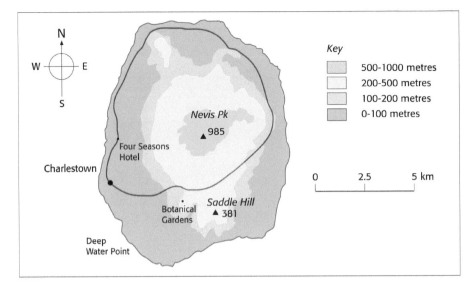

Use Figure 2.6 to answer the following questions.

Exercise

1 What is the shortest distance in metres, to the nearest 100 m, by road between Charlestown and the Four Seasons Hotel?

2 Write the scale of the map in imperial units (miles/inches).

3 Give the six-figure grid reference of the Botanical Gardens.

4 In what direction does Nevis Peak lie from Charlestown?

5 What is the bearing of the Deep Water Port from Nevis Peak?

Describing landforms by reading the contours

A topographical map can be considered as having layers of geographical information on it. The most important foundation information is the shape of the land, or **topography**. This may also be called the **relief** of the land – that is, the height, slope and shape of the landforms.

There are four primary tools for describing the relief of the land:

● referring to **spots heights** and **trigonometrical stations**, which are accurately surveyed points

● noting **contour lines** connecting places of the same height

● calculating the **gradients** of slopes

● drawing **cross-sections** of the map.

Contour lines are brown lines which show the vertical interval (VI) and unit of measurement or 'value' of the lines, e.g. 200 feet or 50 m. More than one vertical interval may be used on a map. For example, an interval of 50 m may be used for land less than 250 m high and an interval of 250 m for land over 250 m high. The map may also have some lighter-coloured form lines and some darker contour lines, which will be explained in the key.

➤ *Figure 2.7 Contour lines on the map of Nevis*

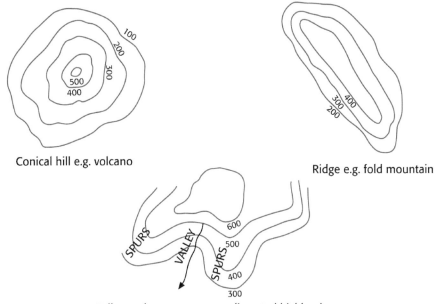

Conical hill e.g. volcano

Ridge e.g. fold mountain

Valley and spurs e.g. many dissected highland areas

Describing landforms

﹡ There are three basic features of any landform: **height**, **slope** and **size**.

Using heights we can identify **highlands** and **lowlands**.

Using the closeness of the contours we can determine the **slope**: contours that are close together show **steep slopes**; lines further apart indicate **gentle slopes**. The scale can be used to determine the size of a landform.

On Caribbean maps we generally look at:

- highlands – volcanic cones, fold mountains (often rugged/steep relief), hills, spurs, cliffs

- lowlands – coastal plains, river valleys, depressions.

You will learn about these landforms in Chapters 4, 6 and 7.

Look at Figure 2.6. The relief of Nevis could be described as follows:

'A main central cone rising to 3 232 feet (grid ref). It is steep-sided (grid square 4096) and has numerous valleys and spurs. An example of a valley is Fountain Ghaut in the north-east of the map. A spur is shown at grid reference 358999.' There are few areas of flat land except along the coast, e.g. Clarke's Estate, grid square 3396.

Cross-sections

✳ A **cross-section** is a profile through a landform. It shows its shape or outline like a silhouette, as if we could cut through the land.

Exercise

Make a small model of a hill (you can use wet sand, soft flour dough, Play Doh, plasticine) of any shape. Include some bumps. Take a ruler and slice through the model. Look at the sliced area – that is a cross-section.

We can complete the same process using the information on a map.

1   Join any two points on the map with a straight line.

2   Put a straight edge of paper along the line.

3   Mark a short line on the paper to indicate the position of each contour line.

4   Write the height of each contour line marked. (You may have to lift up the paper and look for these on the lines.)

5   On a new sheet of paper, draw a horizontal line equal in distance to that between the two map points.

6   Draw a vertical line at the left-hand point.

7   Choose a scale that can represent all the heights on your marked paper (you may want to use the same vertical interval as the map).

8   Put your marked piece of paper along the line and transfer the height information.

9   Draw a vertical line, to scale, from each mark to the correct height and put a dot.

10 Connect the dots by a smooth curve.

11 Label/name any significant natural or human features.

12 Add a title, unit of scale, names of start and end points.

➤ *Figure 2.8 Drawing a cross-section*

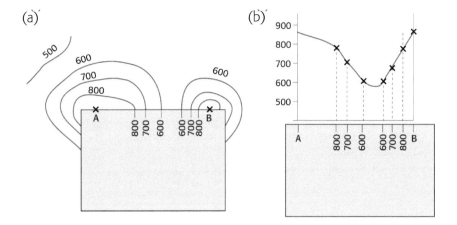

(a)

(b)

Gradients

Gradients indicate the slope of the land. You may have seen a sign on a steep hill: 1:6 – this is the gradient of the slope. Some drivers of cars, buses and trucks may need this information in order to drive safely.

✳ **Gradient** is the relationship between the distance between two points and their difference in height. It may be written as a formula:

$$\frac{\text{Vertical difference in height}}{\text{Horizontal distance}}$$

Example

1 What is the gradient between the summit of Nevis Peak and the coast at Charlestown in Figure 2.7?

a Calculate the difference in height of the two points (use the contour lines).

Height of Nevis Peak is 3 232 feet, height of coast 0 feet. Difference in height 3 232 minus 0 feet = 3 232 feet.

b Measure the distance between the two points (use the line scale).

Distance between Nevis and the coast at Charlestown is 3 miles.

Note: If the heights are in feet, use imperial units to measure the distance.

c Make sure both values are in the **same unit**, e.g. feet.

Multiply 3 miles by 5 280 to convert to feet = 15 840 feet.

d Write the height:distance as a ratio, i.e. 3 232:15 840.

e Cancel down the ratio to 1:*whole number* – in this example 1:5 is the gradient of the slope between Nevis Peak and the coast at Charlestown.

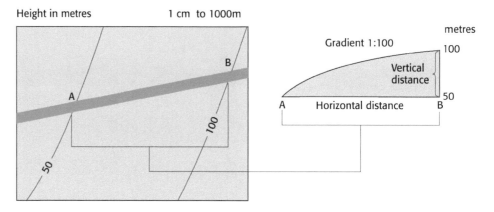

Height in metres   1 cm to 1000m

▲ *Figure 2.9 Calculating the gradient/slope between two points*

Exercise

Use Figure 2.7, the map of Nevis, to answer the following questions.

1 Calculate the average gradient of the slope from Ft George △ 21 feet (326939) to Estate △ 319 feet.

2 Draw a cross-section from N39 (1901 feet) at 403968 eastwards to the sea.

3 Divide the map into highlands and lowlands. Describe the features of each section.

Reading and interpreting conventional symbols

✳ After we have looked at the **scale** and **north arrow** of a map, we then look at the **key** or **legend**.

The key is usually located at the bottom or side of the map. It shows the meanings of all symbols and colours used on the map.

For example: buildings are squares; some are labelled: P for post office, PS for police station, Hosp for hospital, Ch for church, Sch for school. The symbols chosen may vary from map to map so it is important that you look at the key.

Describing drainage

Water on a map is conventionally shaded blue. This will be shown in the key. Seasonal streams may be shown as dashed blue lines, while perennial streams are shown by solid blue lines. Constructed canals are shown as straight blue lines.

Drainage on a map may be 'good' with numerous streams, 'poor' as evidenced by swamps, or 'absent' with the presence of springs indicating underground drainage. (See Chapter 6.)

Good drainage may be described by naming and locating:

- type of pattern – radial, dendritic, trellis
- source, mouth and direction of flow
- specific landforms such as valleys, channels, gorges, river cliffs, deltas, flood plains, meanders.

Poor drainage may be described by naming and locating:

- swamps, marshes and ponds.

Describing vegetation

Vegetation on a map is indicated by symbols, identified in the map key, which refer to the **natural vegetation** (not the crops). In the key, forests may be green shading or tree-shaped symbols (this should not be confused with forest plantation which may indicate cultivated trees); broken forest or trees with scrub as well as mangrove swamps may be indicated. Chapter 9 will help you understand this aspect of maps.

Generally, Caribbean maps show forest on the highest areas, broken forest/scrub on intermediate areas and mangrove swamps on the coast.

You describe vegetation by:

- giving the location and the distribution of the symbol of the particular type
- indicating the type of relief on which it occurs.

Describing land use

Land use includes all the evidence on any map of human activity. This includes all topics covered in Units III and IV in this book.

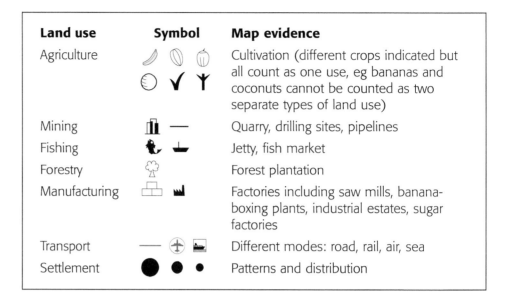

➤ *Table 2.1*

Land use	Symbol	Map evidence
Agriculture		Cultivation (different crops indicated but all count as one use, eg bananas and coconuts cannot be counted as two separate types of land use)
Mining		Quarry, drilling sites, pipelines
Fishing		Jetty, fish market
Forestry		Forest plantation
Manufacturing		Factories including saw mills, banana-boxing plants, industrial estates, sugar factories
Transport		Different modes: road, rail, air, sea
Settlement		Patterns and distribution

key

Named Buildings	
Other Buildings	
Police Station	Pol Sta
Post, Telegraph Office	PT
Hospital	Hosp
Church, Chapel, School	Ch ■ Cha ■ Sch
Court House	CtHo ■ H

▲ *Figure 2.10*
Settlement on a map

key

Main Roads	Bridge
Secondary Roads	Sealed Surface Loose Surface
Other Roads	
Tracks	
Footpaths	
Railways (Light)	

▲ *Figure 2.11*
Communication on a map

Describing settlement

Settlement on a map is indicated by shaded squares. These show where people live and the settlement patterns created by the buildings. The settlement on a map can be described by its:

- distribution – **even** or **uneven**
- density – **dense** or **sparse**
- patterns – **nucleated**, **dispersed** or **linear**.

Describing communications

We can describe the communications on a map in terms of:

- modes of transport – road, rail, airports, seaports
- direction and density of networks for each type
- bridges, fords.

The relationships among patterns of relief, drainage, vegetation, land use, settlement, communications

Seeking relationships between geographical phenomena is a very important skill. (It carries the most marks in the map reading question.) This is the application of all you have learned in the relevant chapters. We look for relationships between relief and all the other aspects.

- Relief influences all other physical and human features:

- relief and settlement – dense settlement on low, flat land; sparse settlement in mountains
- relief and vegetation – forest on hilly areas, mangrove on the coasts
- relief and roads – hairpin bends to travel across steep slopes, roads going along valleys.

● Land use and settlement – settlements may avoid the most useful agricultural land but are located nearby.

● Settlement and communications – settlement is often concentrated at crossroads, or found along a main road.

Note: You must use map evidence only.

➤ *Figure 2.12 Mountains and plains in Central Trinidad*

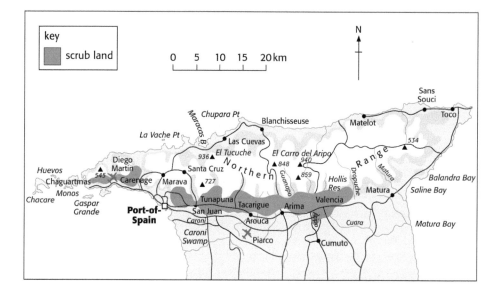

Exercise

Use Figure 2.12 to answer the following questions.

1 Describe the drainage on the map.

2 Describe the distribution of scrub vegetation.

3 Name and locate one area of nucleated settlement, and one area of dispersed settlement.

4 Describe the network of main roads.

Exercise

Use Figure 2.12 to answer the following questions.

1 What is the relationship between relief and communications?

2 Give two reasons for the concentration of settlement to the west of the map.

3 How might the distribution of mangrove be related to drainage?

Go to www.ordnancesurvey.co.uk/freemapsfor11yearolds/PDF_Documents/map_reading_made_easy_peasy for a free booklet on basic map reading.

Photograph interpretation

Photographs show the same geographical information as maps but from many different angles. Photographs may be taken:

● from above like a map – aerial photographs from a plane or satellite

● on the same level as the object

● at an oblique angle to the object – above and looking forward or back.

You may be able to tell the angle by looking at the relative size of the objects and the shading of the photograph.

▲ *Figure 2.13 Different views of the same object*

Photographs may be interpreted by looking at different parts. Figure 2.14 shows how to divide a photograph into foreground, middle ground and background/left, centre and right. Each of these may be described in turn, but generally there is one prominent geographical feature which dominates the centre foreground. However, you may be asked to interpret the photograph, i.e. look for relationships between features.

The photograph in Figure 2.14 may be described as showing:

'... *the Pitons over the Soufrière valley. In the foreground is the steep northern edge of the area, in the middle ground a flat-bottomed valley, while in the background the Pitons dominate the skyline. There is settlement in the low flat area but none on the steeper slopes.*'

➤ *Figure 2.14 Parts of a photograph*

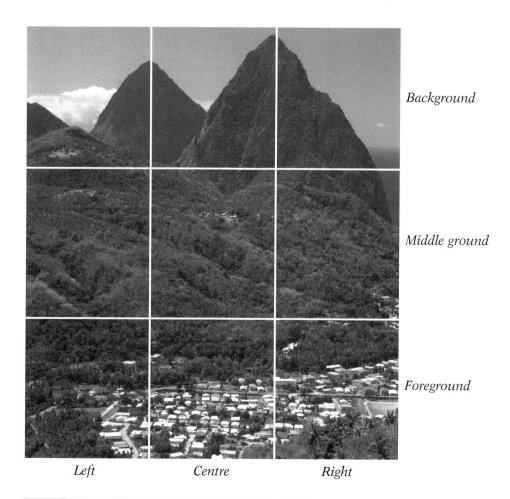

Background

Middle ground

Foreground

Left   *Centre*   *Right*

Exercise

Write two paragraphs describing the impact of the volcanoes on the natural and human features shown in Figure 2.14.

Global positioning

In the past, places had to be located by means of their relative position to other objects. Surveying the land to draw a map depended on human fieldwork, getting data on heights and distances to produce the map. Interestingly, computers and satellites still use the basic triangulation and time/distance methods but applied to the new technology.

Satellites orbiting the Earth can identify any object on the Earth's surface through a computer manipulating the data, to give the latitude, longitude and altitude of that object – Global Postioning Systems. Data is taken from more than one satellite, and trigonometrical relationships, speeds and distances are used to fix the object. (This is essentially the same principle formerly used by sailors who navigated by the stars.)

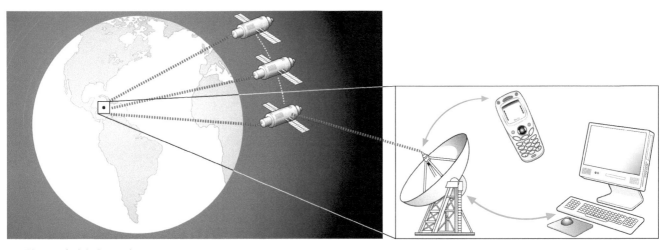

▲ *Figure 2.15 Locating places on the globe*

Latitude and longitude

Lines of **latitude** and **longitude** are similar to grid lines (see page 12). The point of intersection of the two sets of lines can identify a place.

Lines of latitude are drawn from the equator around the middle of the Earth (0° latitude) towards the poles in smaller circles parallel to the equator. So the latitude of a place can be 0–90° north or south of the equator.

Lines of longitude are equally sized north/south circles passing through the poles. The Prime Meridian (0° longitude) passes through Greenwich, England. All other places are 0–180° east or west of the Prime Meridian.

Exercise

Use your atlas or another map to find out the latitude and longitude of the capital of your country.

▼ *Figure 2.16 Caribbean places*

Locating territories in the Caribbean

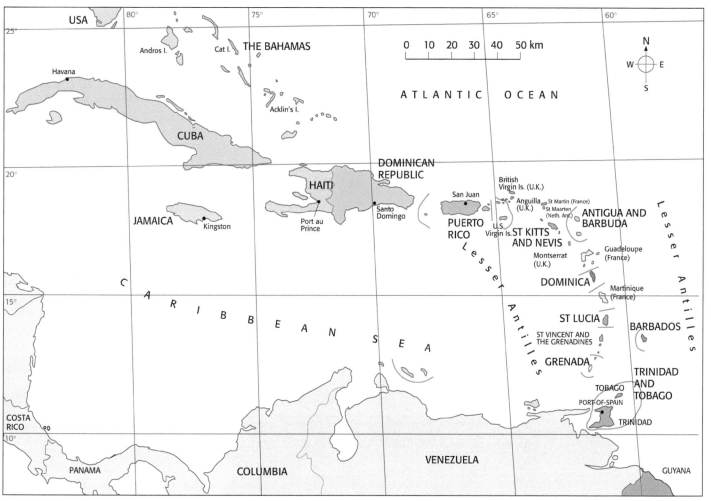

Exercise

Use Figure 2.16 to answer these questions.

1 What are the latitude and longitude of Monserrat?

2 Is Tobago further north than Grenada?

3 What is located at 10° N and 61° W?

Time

The Earth completes one rotation on its axis (through the north/south poles) in 24 hours. One complete rotation is equal to 360° longitude, so we can say that the Earth rotates through 360° in 24 hours. Therefore it rotates through 15° longitude in one hour. Since the Earth is a sphere, we needed to fix time from somewhere. Time is fixed from the Prime Meridian at Greenwich and referred to as Greenwich Mean Time. Places are said to be behind or ahead of Greenwich. We can use longitude to determine time.

If we know the longitude and time in one place and the longitude of another, we can work out the time of the other place.

Example

If it is 2 pm in Bridgetown at 60° W longitude and Santa Barbara, California is at 120° W longitude, then the difference in longitude is 120 minus 60 = 60°.

The Earth turns through 15° in one hour, so 60° = 60 ÷ 15 = 4 hours.

Santa Barbara is further west so it will be 4 hours earlier. (Remember: the sun rises in the east so places further east are ahead of those further west.)

The time in Santa Barbara will be 2 pm minus 4 hours = 10 am.

However, for convenience there are broad time zones: the Caribbean is in the Eastern Standard Time Zone, as is the Eastern USA; Central USA has its own zone, as do the Pacific states. On television, foreign programmes may have the equivalent times at the bottom of the screen. We also notice the difference in time when we watch cricket in Australia, India or South Africa, and basketball in North America, which are in different time zones from the Caribbean.

Exercise

Answer the following questions.

1 If a cricket match starts at 10 am Monday in Kingston, Jamaica at longitude 77° W, what time would it be in Cape Town, South Africa at 18° E?

2 What time is it in Port-of-Spain, Trinidad at 61° W longitude, when a cricket match in Melbourne, Australia at 144° E finishes at 6 pm on Thursday?

3 If the Lakers were playing basketball in Chicago at 88° W at 8 pm, what time would it be in Montego Bay, Jamaica at 78° W?

Sketch maps and diagrams

Sketch maps showing relative location and spatial distribution

A sketch map is a generalised drawing used to illustrate the location or distribution of some geographical feature. It should have a frame/border, title, key, north arrow. A scale is not usually used because sketch maps are, by definition, not drawn to scale. For example, for your SBA you will be required to draw two sketch maps – one to show the location of your study area within your country, and another to show the relative position of the features of your study area itself. You can also use maps to present your data, for example, the relative location of cliffs/beaches along a coastline and wave direction, that you drew in the field.

In order to show any feature you must not only be able to indicate the correct shape of the country or part of territory, but also place the required area correctly and in the right size or extent.

For example, if you are drawing a sketch map to show location of commercial arable farming in Guyana.

Step 1

Draw a frame and put the title at top. Put a north arrow and allocate space for a key.

Step 2

Put in the coastline in blue or black. (Since Guyana is part of a continent, make sure that you put in the coastline right up to the frame.) Label the water – Atlantic Ocean.

Step 3

In red, roughly sketch in the land boundaries of Guyana to west, east and south. You do NOT need to put in every bend, just the general elongate/double-rectangular shape. Write in your key: Red – country border.

Hint: Most countries can be sketched as general shapes, e.g. Trinidad as a rectangle with points, Jamaica's car-like shape and so on.

Step 4

Put in the rivers (in blue) or towns (red dots) that mark the start and end of the farming area, e.g. Essequibo, Demerara and Berbice rivers and/or Georgetown and New Amsterdam.

Step 5

Estimate how far from the coast on your map farming is carried out and shade the area green. Name a town or village in the area, e.g. Mahaicony. Add to your key: Green – commercial arable area of Guyana.

Note: You need to show related features on your map to correctly show its distribution and extent.

➤ *Fig 2.17 Sketch map of Guyana, showing commercial farming areas*

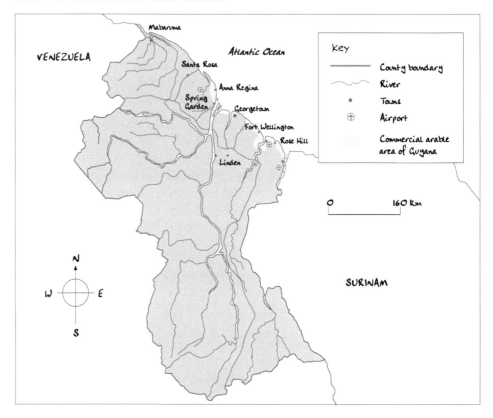

Diagrams to illustrate geographical features

Diagrams can be drawn in many different forms: plan (like a map), cross-section, front elevation or even block. Cross-sectional diagrams are often the most useful to show the relative position of geographical features.

You could draw a well-labelled diagram to show the chosen features of a composite volcanic cone as follows.

Step 1

Decide what type of diagram could best show these – perhaps a block diagram, but these are time-consuming to draw, so a cross section may be used.

Step 2

Identify the essential features which are to be shown. These are the shape of the cone, vent, ash and lava layers.

Step 3

Draw the correct shape of cone above the ground.

Step 4

Use a key to show layers of ash and lava and vent. Label them.

➤ *Fig 2.18 Cross-section of a composite volcanic cone*

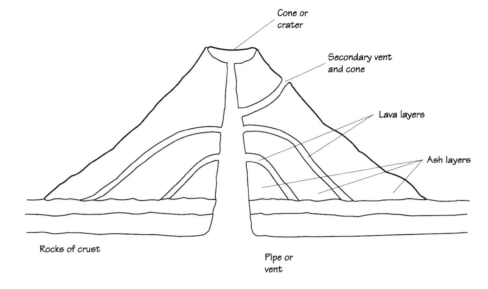

Notice that the labels are identifying the parts of a composite cone: not just any volcano.

If we were asked to draw diagrams showing the formation of ox bow lakes we may use diagrams in plan form, which show the processes at work. In the field, we often draw simple 3-D representations of the main geographical features of the landscape. This is not artwork: this is graphically showing the main features you are studying. Some people may use photographs, but sketching has the advantage of allowing you to show only those features you are interested in.

Step 1

Draw a frame and write a title for your diagram.

Create a 4-sided card frame to look through. Centre your frame on the feature that you want to highlight.

Step 2

Put in with light lines the main outline of the feature, e.g. if you are investigating the relationship between farming and forests in a mountainous area of Trinidad, the main feature you'll wish to show will be the farms located on steep slopes surrounded by forested areas. So sketch in the sides of the valleys and hill slopes. Farms could be indicated by their boundaries; similarly forested areas could be shown by simple tree shapes.

Step 3

Show different crops with symbols, e.g. terracing, plough marks, tree crops, field crops.

Fill in and label other features – roads, rivers, farm house, other settlements.

Statistical diagrams

This section is very important for presenting the data in your School-Based Assessment field study (Chapter 3). You are required to present your data in three different forms and then to interpret these in an integrated presentation.

Once you have collected your data, you have to decide which diagrams would best represent it visually. Diagrams and tables summarise the data and often allow you to see patterns and trends. For example, if your data is continuous (temperatures, tourist arrivals, oil production) and varying over time, then a line graph could be used.

Line graphs

1 Draw a horizontal line to represent time intervals.

2 Draw a vertical line at the left end.

3 Look at the highest and lowest values to be represented.

4 Decide on a scale that will allow all points to be plotted, e.g. 1 cm represents 5°C.

5 Mark off the scale on the vertical axis.

6 Mark off the time intervals on the horizontal axis.

7 Indicate the unit used on each axis.

8 Plot the value against the corresponding time and place an X at that point.

9 Join all the points using a ruler.

10 Add a title to your graph.

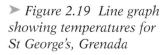 *Figure 2.19 Line graph showing temperatures for St George's, Grenada*

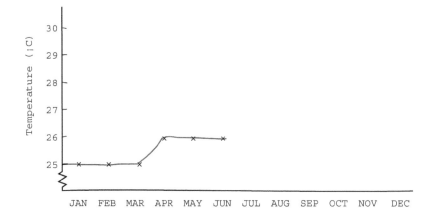

Bar charts

Bar charts are useful in showing data that is discrete and not connected, for example, rainfall, or population of Caribbean capitals.

To draw a bar chart, follow steps 1–7 as for line graphs and then draw bars from the horizontal axis up to the corresponding value on the vertical axis.

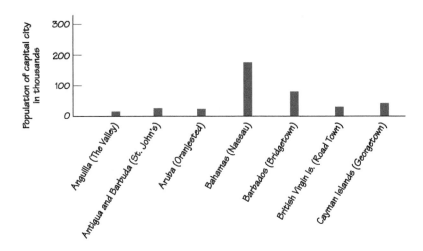

➤ *Figure 2.20 Population of Caribbean capitals*

Divided circles

Divided circles are sometimes called **pie charts** because they represent data in 'slices' or sectors of the circle. This is the best way of representing proportional data or items that add up to 100 per cent. The whole circle of 360° represents 100 per cent, so 1 per cent is represented by 3.6°.

➤ *Figure 2.21 Drawing divided circles*

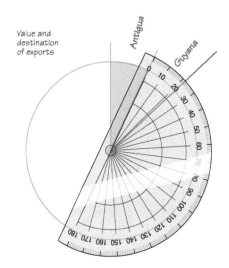

Country to which goods are exported	Value of exports (US$)	Percentage of total exports	Calculation	Angle
Antigua	245000	7	360 × 0.07	25.2
Guyana	210000	6	360 × 0.06	21.6
Jamaica	560000	16	360 × 0.16	57.6
UK	12950000	37	360 × 0.37	133.2
USA	280000	8	360 × 0.08	28.8
Other	910000	26	360 × 0.26	93.6
Total	3500000	100	Check	360

from Dominica 1994

Interpreting line and bar graphs and divided circles

In addition to drawing graphs, you may be asked to interpret them.

- Be clear about what the diagram is showing – the main variables.

- Look at the largest and smallest values, i.e. the range of values.

- Note any exceptions.

- Look for a relationship between the two units – is one increasing with the other? Are the bars clustered on any values?

- Seek patterns or trends in the data.

Exercise

You have completed your School-Based Assessment on rural to urban migration in Port-of-Spain, Trinidad. Using a questionnaire, you have collected the following data:

Answers to Q2: *Where did you live before moving to Port-of-Spain?*

Diego Martin 6; Arima 3; San Juan 14; Chaguanas 7

Answers to Q3: *What was your reason for moving?*

To find work 15; To join family 5; To be close to entertainment 10

Answers to Q4: *Do you have your own car?*

Yes 20; No 10

Decide how to represent each of these items and draw the appropriate diagram. Check that each diagram has a title, frame and key.

Write a brief paragraph interpreting the data shown.

Interpreting maps

You can also be asked interpret **dot maps**, **choropleth maps** and **isopleth maps**, although you are not required to draw them.

Dot maps
Look at Figure 2.22a on page 37, which is a dot map showing the location of fast food restaurants in Barbados. Each dot represents one restaurant. By looking at the distribution of the dots we can see that most of the fast food places are concentrated around Bridgetown and along the coast. The interior has fewer restaurants. Why do you think they are concentrated in these areas? (Hint: tourism, urban areas.)

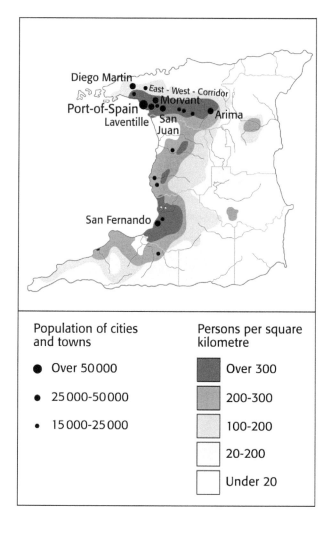

Choropleth maps

These shaded maps show the density of the feature. Figure 2.22b shows a choropleth map of the population of Trinidad.

In interpreting the map you need to identify the highest and lowest values. Look for any other variables associated with each, e.g. near the coast, near a factory or town.

Isopleth maps

These are maps showing lines of equal value (isolines). A contour map is a type of isopleth map. Many geographical variables can be mapped using isolines. Figure 2.22c shows an isopleth map of hurricane frequency in the Caribbean over 20 years.

◀ ▼ *Figure 2.22a–c A dot map (left), a choropleth map (below left) and an isopleth map (below right)*

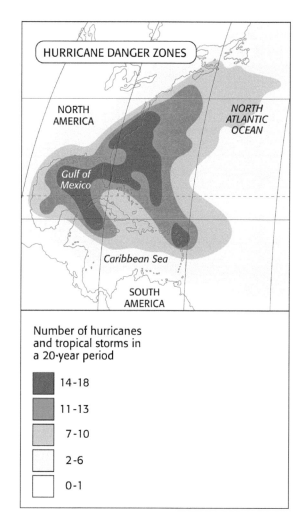

CHAPTER 3

The School-Based Assessment

By the end of the chapter students should be able to:

- **identify and define a problem** or situation relating to the syllabus

- **apply skills, knowledge** and **principles of geography** to the field study

- **collect**, **record** and **present information** based on field study work on at least *one* chosen topic.

Field study

When you begin doing CSEC-level work, the **School-Based Assessment (SBA)** should be viewed as the fun part of your studies. The Geography SBA requires you to go out in the field to observe, investigate and identify what you have been reading and learning in class and to record your findings. Field study is at the heart of geographical investigation. Geography is concerned with natural and human phenomena, which are best studied in the field.

Your School-Based Assessment (SBA) is a field study report. 'The field study is the School-Based Assessment component of the Geography syllabus and is included in the General Proficiency Examination.' Caribbean Examination Council, CSEC Geography Syllabus 2005

Field study is the primary investigation, observation and measurement of things in the real world. The report of a field study is *not* an assignment for which you go to the library or internet to do research. It has to be a report you write from observation, measurements and recordings that *you* have collected in the field.

While preparing for the CSEC Geography examination, students can make distinctive contributions to environmental understanding through field study. The SBA component of the CSEC examination is actually preparing students for more advanced research at the tertiary level of study and later on in the management of the environment.

CXC recommends that you start the SBA at the end of the first year. When you submit your SBA to your teacher before or on the specified date, you can be confident of your exam performance. The SBA ensures that you have some:

- knowledge of one topic
- practical skills
- analytical skills.

A good SBA (20% total assessment) can contribute to a Grade 1 or 2.

Choosing a topic

This is an essential part of doing the SBA. You should have completed studying the topic you intend to do before you start the SBA.

Think about the following before you start:

- What do you want to research?
- Where will you collect the information?
- How will you collect the information?

(See Figure 3.2 on page 42.)

Exercise

Make a list of all the problems (as stated in the syllabus) that you can identify in your community or town which interest you. You may like to consider these:

- Erosion of a section of a beach or cliff side
- Solid waste pollution along a stretch of beach
- Landslide along a road
- Traffic congestion in a city
- Squatter settlements
- Effect of local climate on a community
- Effects of globalisation on an economic activity in a particular community

The **topic** of the SBA is the general topic taken from the syllabus. It may be taken from any one of three units, as illustrated in Table 3.1.

Units	Topics	Sample titles/questions
Natural systems	Volcanoes, weathering, mass movement, wave and river processes, coasts, river channels, limestone and karst landscape, weather, components of an ecosystem, characteristics of vegetation and soil	1 How does longshore drift contribute to beach development in a section of Speyside, Tobago? 2 What factors are contributing to beach erosion at Carlisle Bay, Barbados? 3 What processes contribute to the weathering of rocks on the cliff face along the Gordon Town Main Road, Jamaica?
Human systems	Population distribution and movement, urbanisation, economic activities – fishing, agriculture, forestry, mining, food processing, garment manufacturing, tourism Challenges in economic activities – globalisation, technology, sustainability and marketing (Caribbean Single Market and Economy/CSME)	1 What factors have contributed to the growth of population in Portmore, Jamaica over the past 20 years? 2 What factors have contributed to the location of the food processing plant in Arima, Trinidad? 3 What challenges does the Jalousie Resort in Soufrière, St Lucia, face?
Human–Environment systems	Impact of natural hazards, pollution, global warming, deforestation, coral reef destruction	1 Does deforestation contribute to soil erosion in sections of Cumberland, St Vincent? 2 What effect did Hurricane Ivan have on St George's, Grenada? 3 What are the types of pollution that affect the town of San Fernando, Trinidad?

▲ *Table 3.1 Some suggested SBA topics*

Important: Your general topic of research *must* be taken from the syllabus. Agriculture, for example, is on the syllabus; however, the syllabus objectives clearly state the specific area of agriculture that is to be covered. So if pig farming is not in the specific objectives, you should *not* do research on pig farming.

➤ *Figure 3.1 Choosing a topic*

Note that the **title** says specifically what you intend to study. In the last column of Table 3.1 there is a list of suggested titles. You may reword these to match a problem or situation in a location of your choice.

'Human–Environment systems' is new to the syllabus. It is an area where you will probably find it easy to come up with a title and location. A few are presented here for you:

1 What measures are being taken to conserve the forest along the Fairy Glade Trail in New Castle, Jamaica? (*Any other trail within a forest reserve that is threatened may be observed.*)

2 Why do people in Guava Ridge, St Andrew, Jamaica, continually clear forest vegetation? (*Or any other community you know.*)

3 Do school grounds suffer from air pollution from passing vehicles?

4 What effect has the removal of natural vegetation from areas around the coastline (*place here the name of a stretch of coastline of your choice*) had on offshore reefs?

5 What measures are being carried out to prevent further removal of natural vegetation in Buccament, St Vincent?

6 What role does the natural vegetation play in the ecosystem of Holywell Park in Jamaica?

7 Is industrial pollution a problem in Marabella, Trinidad?

8 What type of pollution is affecting the Black River system in Maggotty, Jamaica?

Conducting the field enquiry

> ➤ *Figure 3.2 Plan of action*

Choose a topic

- Your own observation (field trip)
- Discussion in class or with people outside of school
- Teacher's guidance or suggestion
- Reading

Narrow the topic

- What and how much information is needed?
- How will the information be collected?
- How will the information be presented (in three different ways)?

Write an aim

- What do you want to achieve?
- Choose a site for fieldwork.
- Where will you go?
- Is information available?
- Is it safe?

After drafting a plan, write the outline on a **strategy sheet** which will be given to you by your teacher.

- Your teacher must first agree to your plan with you and guide you through the plan.

- You must fill in the strategy sheet and return it to your teacher before starting your research. The teacher needs to guide you to ensure that you are following the correct procedure.

- You must read before going out to gather data. Reading about your topic of choice will help you to decide on the problem or question and to know what to look for. Remember to make a note of all the materials you read as you will need information like authors' names, publishers, and dates and places of publication for your bibliography.

Getting ready for the field trip

You will need:

- notepad, pencil, pen, map or route plan

- measuring equipment, for example compass, GPS etc. (depending on topic)

- record sheets, interview schedule, questionnaire (depending on topic)

- comfortable shoes, clothing, raincoat, hat, sunscreen and plenty of drinking water.

Data can be collected in groups but each student must analyse and present his or her own findings. Remember to make careful observations and keep records in the field. Record the time of collecting data and any unexpected changes in events.

▲ *Figure 3.3 Field trip*

Presenting your findings

1 The **table of contents (1 mark)** should be clear and show the page numbers of the following:

Table of contents	Page number
Aim	
Location of study	
Methodology	
Presentation of data (illustrations, quality of data, analysis and discussion)	
Conclusion	
Bibliography	

Your pages should be properly numbered. That is, the headings on the contents page and the corresponding numbers must be consistent throughout the pages. There should be no subheadings.

2 The **title** of your SBA is the specific problem or question that is related to the general topic. It is written on the cover page, along with your name and year of examination. The field study can cover only a small area and give a small account of the bigger picture. The question allows you to come up with answers while carrying out the research. You will not be expected to research a complete topic, but you must choose an aspect of the syllabus that is manageable. For example, you cannot look at agriculture or

tourism in your country. Neither can you study an entire river channel. Instead you can study a farm, or tourism activities in a small town.

Topic: Wave action

Sample: 'How did wave action contribute to the development of coastal landforms along the _____ (name a stretch of coastline and the territory)?'

Topic: Hazards

Sample: 'What is the effect of Hurricane Ivan on the community of _____ (name the community and territory)?'

Topic: Vegetation and Soils

Sample: 'What role does the natural vegetation play in the ecosystem of Holywell Park in Jamaica'?

The third title above will be the focus of examples throughout this section.

Note that the title says specifically what you intend to study and where. Discuss this with your teacher **before** you start.

3 The **aim** of the SBA **(2 marks)** is a statement that says what it is that you will achieve in answering your question. The statement should have at least two descriptive words showing how you will collect the data and how you will present this data in order to answer the question or problem. You include your title in the aim in the form of a question.

The aim of the study uses descriptive words such as: *identify, determine, find out, examine, investigate, observe* (words that can describe ways of collecting information); *describe, explain, illustrate, discuss, compare* (words that can describe ways of presenting). The aim helps:

- in determining how the data will be **collected**; for example, an investigation may require certain instruments or tools to carry out experiments

- in determining how the information will be **presented**; for example, if the aim states 'to illustrate', then this would imply the use of diagrams and charts where applicable.

An aim for our example title may go like this:

Title: 'What role does the natural vegetation play in the ecosystem of Holywell Park in Jamaica?'

Aim: 'To examine and illustrate the role the natural vegetation plays in the ecosystem of Holywell Park in Jamaica.'

When going out in the field to collect data the aim is 'to examine'. The techniques you use should allow you to observe the area keenly and to come up with findings (this will be discussed in detail in point 5 below), and the findings should be 'illustrated' in your presentation.

4 The **location** of the field study **(4 marks)** is the place where the research is carried out. In this section you will use some of the skills learned in Chapter 2 to present your maps. The location of your study must be a small area, for example a stretch of beach, section of a town or even your school compound. Your location must be illustrated on two sketch maps: one of the country in which the study area is located and one detailed sketch map of the actual location. The sketch maps must have all the necessary requirements of a map (title, key, north point, border) and should be neat but not drawn to scale. If the map is copied from another map this should be indicated, and it should be sketched, not traced.

The maps should show:

● your study area (use a fine red pen and mark the study area with an X)

● your study area in relation to other features such as roads, rivers, settlement, topography, land use and any other feature important to your study.

Use of techniques such as grid reference, contour lines and other topographical symbols will enhance your maps and give extra credit. Remember, the skills and principles you learn must be applied to your study. Your teacher may be able to lend you topographical maps or you could look at or borrow them from your local survey department – and remember to use your atlas.

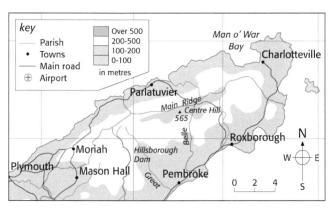

▲ *Figure 3.4 A topographical map – location of study area*

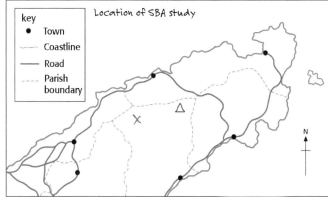

▲ *Figure 3.5 A sketch map – features of study area*

5 The **methodology (4 marks)** is very important in stating exactly what you did to collect your data, i.e. how, when, where, and providing answers to your title question. **Remember:** in the aim you stated what you were going to do to address the problem/answer the question, but the aim doesn't provide an opportunity to explain the instruments used. Your SBA must give a brief description of *how*, *when* and *where* data was collected for the study.

- **How (2 marks):** this is to explain the procedures you used to get information. **Remember:** this is field research and though reading is very important, practical activities must be done to get answers for yourself. Here are some methods you can use depending on your title:
  - *Questionnaire:* a set of questions with answer choices designed to get a desired response. This method rules out unnecessary data manipulation and addresses the specific time period you may want to examine.
  - *Interview schedule:* a list of questions to ask an interviewee.
  - *Record sheet:* a prepared table to record readings from observation or from instrument measurements. For example, when recording wave frequency you will need to have a prepared record sheet to record counts.
  - *Quadrat:* a square subdivided into equal units to form a grid. Quadrats are good for measuring density and distribution, especially of species in an area. This allows you to interpret or analyse a sample of the wider area. This method is good for doing a topic on ecosystems.
  - *Line transect:* a line drawn with a cord for about 30 metres and marked at intervals. Good for studying changes along a gradient, for example changes in pebble size from foreshore to backshore up a beach, or changes in vegetation along a slope.
  - *Maps:* topographical maps provide base information for your sketch maps.

- Secondary data, textbooks, magazines and any written material that can help to interpret, analyse and illustrate information collected in the field. **Remember:** you must not plagiarise, and you must make reference in your bibliography to any written material that you use.

- **When (1 mark):** give a clear indication of the date or dates when information was collected. In some instances the time of day is important, so if you are working on a title that is affected by the time of day then state this.

▲ *Figure 3.6 Using a quadrat*

- **Where (1 mark):** include a brief statement describing the geographical coordinates of the study area, for example, 'Holywell is in north-east St Andrew just 1 km south of the border of St Andrew and Portland at an elevation of approximately 1 500 m.' Make sure that this location and the features described can be identified on your maps.

6 **Presentation of data (18 marks)** is a vital part of your field study. You should refer to your diagrams and discuss them – your illustrations and their analysis should be **integrated** into your work. Now you are back from the field, you have all the information and must now put it on paper. Look back at your aim – your presentation should relate to your aim. So if your aim was to '*examine* and *illustrate* the role the natural vegetation plays in the ecosystem of Holywell Park in Jamaica', your presentation should describe that role using various illustrations. The written account must be well organised and follow a sequence. Always introduce the topic using a definition or statement in the first paragraph. Then say how your area of study relates to this definition by inserting the data. So if your general topic is 'vegetation and soil', you must say *how* vegetation affects the ecosystem of your study area. The data you collect must relate to the general topic in answering the question: 'What role does the natural vegetation play in the ecosystem of Holywell Park in Jamaica?' Suppose you find that one role is maintaining soil moisture: how could you present this? One way would be to create a table – you then analyse and discuss the table. You may find that another role is supporting animal life. This sort of information would have been collected from your quadrat sampling. The information may now be presented in a graph. The interpretation of the graph may show that as vegetation increases, animals and other organisms increase in number, and where vegetation is sparse, for example in a clearing, the number of organisms decreases.

Do...	Don't...
Write clear and relevant information.	Plagiarise or photocopy material from books.
Include illustrations of your study area – pictures, diagrams, tables.	Use generic pictures or tables from textbooks (if you use diagrams from textbooks they should be adapted to your study and this should be indicated in your work).
Label all pictures and diagrams as figures and tables and charts.	Leave illustrations unlabelled.
Integrate illustrations in the text and refer to them in your text.	Put illustrations, especially photos, at the end of your study or in the appendix.
Stick within the word limit of 1 650 words (approximately 13–15 pages).	Include unnecessary material and pictures that 'pad out' the SBA and take it over the page limit (you will be penalised 10% of your score for exceeding the word limit).

▲ *Table 3.2 Dos and don'ts in the presentation of data*

Many of the techniques you should use in presenting your data can be found in Chapter 2. These include:

- **different types of maps:** topographical, and dot and line

- **charts and graphs:** bar graphs, line graphs and pie charts – remember that the segments in any of these must be representative of your data; for example, 10% of data is represented in the circle by a segment of 36°

- **diagrams:** block diagrams and sketch drawings – diagrams must be properly labelled and well integrated in text

- **photographs:** use only photographs that were taken at the time of study – do not use photocopied pictures.

There are 8 marks for 'diagrams' and 10 marks for their discussion and analysis.

7 The **conclusion (6 marks)** should be a brief paragraph summarising the main points that were made and must be related to the aim. The conclusion should answer the question asked in the title. It may also include suggestions for solving the problem and/or state the implications if the problem is ignored.

➤ *Figure 3.7 Example of a bibliography*

BIBLIOGRAPHY

1 Briggs, David and Smithson, Peter. *Fundamentals of Physical Geography.* New Jersey: Rowman & Littlefield. 1994

2 Gentles, Marolyn Lucy and Ottley, Jeanette. *Longman Geography for CSEC.* Harlow: Pearson Education. 2005

3 Nagle, Garrett. *Geography Through Diagrams.* Oxford: Oxford University Press. 1998

4 Poxon, E.M. and J.D. *Photo Mapwork for the Caribbean.* Edinburgh: Ginn. 1968

5 *Longman Atlas.* Harlow: Longman UK. 1998

6 *Hutchinson Dictionary of Geography.* London: Helicon. 1997

7 'Today's Weather Report and Forecast'. *The Daily Gleaner.* Saturday 12th February 2005.

---

8 The **bibliography (1 mark)** is a list of all sources, recorded in alphabetical order of authors. Include the name of the publisher and date of publication. Try to read at least three sources apart from your textbook. You will not get substantial information from one book only. If you use the internet you should cite the websites you have visited.

9 The **appendix** should include samples of any questionnaires, interview schedules or record sheets used (do not put photos here).

Note: A maximum of 4 marks can be awarded for correct grammar and use of geographical terms. There is also a penalty of -4 marks for exceeding the word limit.

A completed SBA should show all the skills, knowledge and principles of geography that are necessary for the topic you chose. **(Total marks = 40.)**

Exam practice questions

Practical skills

Paper 1: Multiple choice

Use the map in Figure P.1 to answer the following questions.

1   In which direction does the River Caribbee flow?
    A South east
    B North west
    C West
    D East

2   The distance in a straight line from Unity Peak to Independence Town is:
    A 3.0 km
    B 2.5 km
    C 1.7 km
    D 2.7 km

3   Which landform is shown at point C?
    A Mountain
    B Depression
    C Valley
    D Knoll

4   The site of Independence Town can be best described as:
    A The coastal plain
    B At an estuary
    C On the beach
    D On lowlands

 Figure P.1

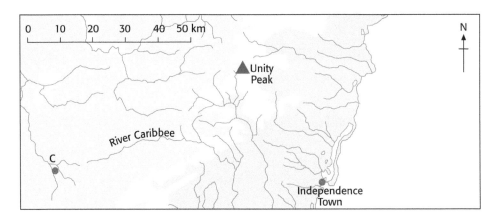

➤ *Figure P.2*

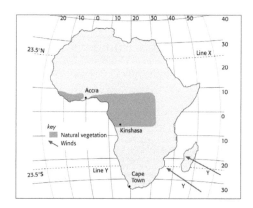

5   The approximate latitude of Cape Town on the above map is:
    A 34°S
    B 34°N
    C 18°W
    D 18°E

6   The point marked X on the map is the:
    A Equator
    B Tropic of Cancer
    C Tropic of Capricorn
    D Prime Meridian

7   If the time in Accra (0°) is 0900 hours (9.00 am) what will the time
    be in Kinshasa (15°E)?
    A 8.00 am
    B 10.00 am
    C 8.00 pm
    D 10.00 pm

8   How would a scale of 4 centimetres to 1 kilometre be shown as a
    representative fraction?
    A 1:2 500
    B 1:50 000
    C 1:25 000
    D 1:100 000

➤ *Figure P.3*

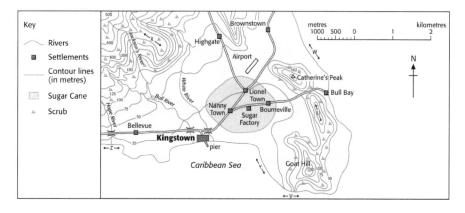

Use Figure P.3 to answer the following questions.

9 Which village lies approximately north west of Bull Bay?
   A Bourneville
   B Kingstown
   C Bellevue
   D Brownstown

10 What is the distance by road from Brownstown to Lionel Town?
   A 1.9 km
   B 1.6 km
   C 2.2 km
   D 2.9 km

11 What is the approximate height above sea level of Highgate?
   A 50 metres
   B 75 metres
   C 65 metres
   D 85 metres

12 Which of these statements best describes the location of Lionel Town? It lies approximately:
   A East south east of Bull Bay
   B South east of Bourneville
   C North east of Nanny town
   D South east of Brownstown

13 In which of these areas is the land steepest?
   A Just north of Kingstown
   B Along the coast near Bull Bay
   C In the extreme north west corner of the map
   D Just south of the airport

14 Which of the following best describes the relief of the area between Belleview and Kingstown?
   A Low-lying and steeply sloping
   B Low-lying and gently sloping
   C Absolutely flat
   D High and steeply sloping

15 The lines on a map that join places of equal height above sea level are called:
   A Relief lines
   B Isopleths
   C Contour lines
   D Grid lines

Total 28 marks

Paper 2 Map reading

The map reading question is compulsory in the exam.

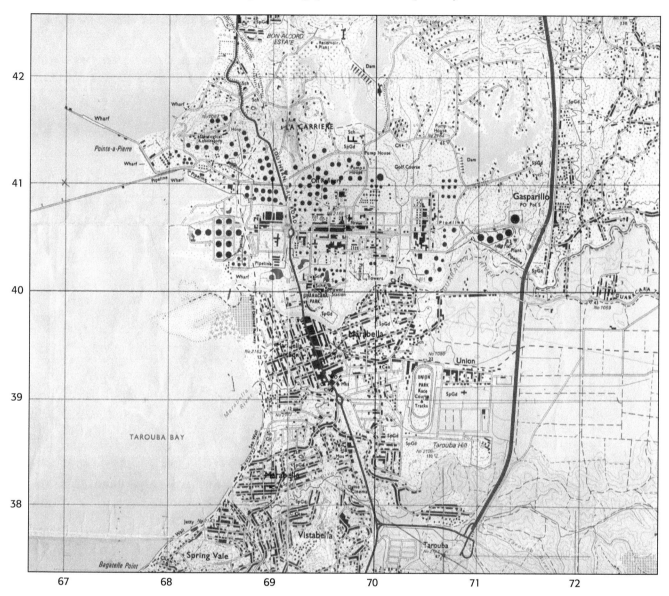

▲ *Figure P.4 A section of the map of San Fernando*

Note: In the actual examination you will be given the whole sheet of an area.

1 Study the map extract of San Fernando, Trinidad, on a scale of 1:25 000, then answer the following questions.

(a) (i) What is the direction of Guaracara Park (square 6939) from Union Park (square 7039)? (1 mark)

(ii) What is the six-figure grid reference of the Trigonometrical Station No. 1050, north of Union Park? (1 mark)

(iii) What feature is located at 675373? (1 mark)

(iv) What is the length of the section of the Solomon Hochoy Highway from the bridge at 714399 southwards to the junction at 710378 (in km to nearest hundred metres)? (2 marks)

(v) Calculate the average gradient of the slope between Trigonometrical Station No. 1050 at height 72 feet and Trigonometrical Station No. 163 (713405) at height 53 feet. (3 marks)

(b) What is the main economic activity in square 7138? (1 mark)

(c) (i) Using map evidence only, name two functions of Marabella (square 6939). (4 marks)

(ii) Name the settlement pattern shown in square 7237. (1 mark)

(d) Describe the relief of the land in square 7138. (4 marks)

(e) Using map evidence to locate an area where coastal pollution may be a problem, explain why it may be a problem in the area chosen. (4 marks)

(d) Describe two potential hazards in the event of an earthquake in the area bounded by eastings 68 and 70 and northings 40 and 42. (6 marks)

Total 28 marks

CHAPTER 4

Internal forces

By the end of the chapter students should be able to:

- define **crustal plates**
- name and locate the **Caribbean plate** and **adjacent plates**
- distinguish among **convergent**, **divergent** and **transform plate margins**
- explain the **formation** and **distribution** of **earthquakes, volcanoes** and **fold mountains**
- explain the formation of **extrusive** and **intrusive volcanic features** and **how these landforms change over time**.

➤ *Figure 4.1 Structure of the Earth*

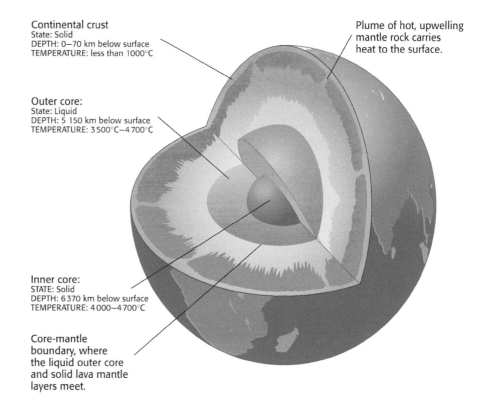

Continental crust
State: Solid
DEPTH: 0—70 km below surface
TEMPERATURE: less than 1000°C

Plume of hot, upwelling mantle rock carries heat to the surface.

Outer core:
State: Liquid
DEPTH: 5 150 km below surface
TEMPERATURE: 3 500°C—4 700°C

Inner core:
STATE: Solid
DEPTH: 6 370 km below surface
TEMPERATURE: 4 000—4 700°C

Core-mantle boundary, where the liquid outer core and solid lava mantle layers meet.

✳ The Earth is a sphere made up of three distinct layers: the innermost **core**, surrounded by the **mantle**, with the **crust** on the surface. These layers are shown in Figure 4.1.

Crustal plates

The theory of plate tectonics suggests that the Earth's **crust** is made up of two layers. The crust is the relatively thin, outer surface of the Earth. The theory states that the lower crust is made of semi-rigid slabs (plates) with the continents and oceans on top of them. There are no spaces between the plates. Some plates have great thickness of old continental material and are referred to as **continental plates**; others have thin layers of rock under the oceans and are called **oceanic plates**. As the plates move, they carry the continents with them.

The theory further suggests that plates move as a result of movements in the molten mantle. As they move they form three types of boundaries with each other:

● **convergent margins** where plates come together

● **divergent margins** where plates move apart

● **transform margins** where they slide alongside each other.

Movement along the plate margins causes earthquakes, volcanoes and fold mountains. These create hazards to Caribbean life and property (Chapter 14). The centres of the plates are generally rigid and stable.

✳ **Crustal plates** are irregular-shaped slabs of material in the lower crust, of varied size. The continents and oceans lie on these plates. Plates float over the semi-liquid material of the mantle. Most plates have both continental and oceanic materials. There are seven large, major plates and many smaller ones.

Exercise

Use an atlas to find a world map showing all the plates.

1 From the map, list all the big plates and some of the small plates.

2 Is the Caribbean plate a big plate or a small plate?

3 Which plates lie next to the Caribbean plate?

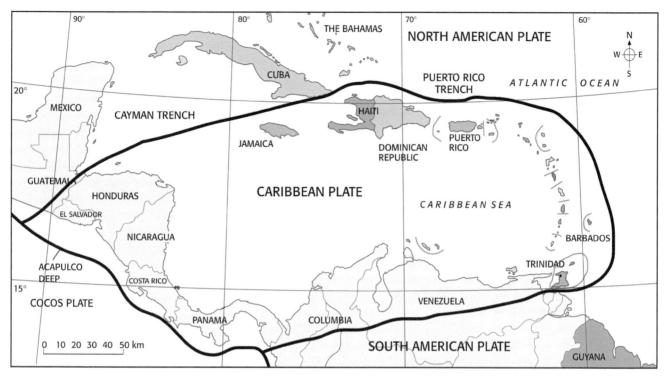

Figure 4.2 Structure of the Caribbean

Exercise

Draw a map of the Caribbean and the adjacent plates (refer to Figure 4.2).

1 Show each type of margin in a different colour.

2 Stick the map onto a piece of light cardboard.

3 Cut the card at the plate margins.

4 Separate the pieces and then reassemble them.

Plate margins

✳ **Plate margins**, or boundaries, may occur under land or sea. There are no spaces between the plates.

The type of margin between plates depends on the movements in the mantle under the boundary, and their relative motion.

Convergent margins

At **convergent margins** plates are moving **towards each other**. Plate material is being destroyed as one plate goes under (subducts) the other. These are also called **destructive margins**. The subducting plate is melted as it goes back into the hot mantle. Convergent margins are marked by earthquakes, fold mountains and volcanoes. The area of subduction is called the **subduction zone**. Subduction zones are associated with continental plates (e.g. North American plate) meeting oceanic plates (e.g. Pacific plate).

Convergent margins in the Caribbean

In the eastern Caribbean, the oceanic Caribbean plate meets the continental North American plate, which is moving back into the mantle. Magma is released through the Caribbean plate, forming volcanoes from Montserrat to Grenada.

Barbados lies east of the subduction zone and is composed of light sedimentary rocks which are being folded as the heavier plate materials are subducted.

To the west of the Caribbean plate, the oceanic Cocos plate is pushing under it, forming another convergent margin. This boundary is marked by the fold mountains and volcanoes of Central America which are also on the Caribbean plate. This is part of the long convergent margin between the Pacific plate and the American plates, creating the Rocky Mountains and the Andes fold mountain ranges with their active volcanoes. These boundaries are shown on Figure 4.3.

➤ *Figure 4.3 Caribbean convergent margins*

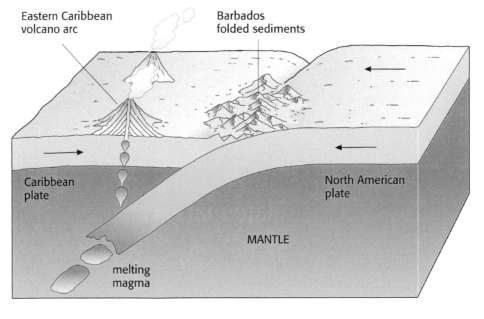

Divergent margins

At **divergent margins**, plates **move away from each other** and new plate material is added. These are also known as **constructive plate margins**. This new material is magma upwelling from the mantle. Continents on the separating plates may be broken apart at these margins, which are marked by earthquakes and volcanic activity.

In the past, Africa and the Americas formed one continent. It was separated by upwelling in the Mid-Atlantic Ocean to create the separate continents. This area is marked by the Mid-Atlantic Ridge. At present, the East African rift valley is thought to be the birthplace of a new sea, breaking up continental Africa.

There are divergent margins in all the ocean basins. The Mid-Atlantic Ridge is the largest divergent plate margin on the planet.

Divergent margins in the Caribbean
In the Caribbean there are no large divergent margins and only one possible area of upwelling, around the Cayman Islands.

▼ *Figure 4.4 Divergent margin in East Africa*

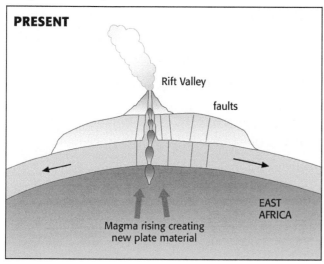

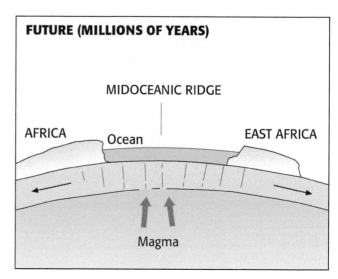

Transform margins

At **transform margins,** plates **slide horizontally** past each other. There is little or no vertical displacement between these plates. There are many such margins throughout the world.

One significant transform margin is the one marked by the San Andreas Fault in California, USA. This area of California, along the coast, is subject to major earthquakes, but it is still a densely populated area. Here the very small Juan de Fucas plate, which includes the Californian peninsula, is sliding north-westwards while the larger neighbouring North American plate is moving in a westerly direction.

Transform margins in the Caribbean
The eastward-moving Caribbean plate has transform boundaries to the north and south, alongside the larger westward-moving North American and South American plates (see Figure 4.5).

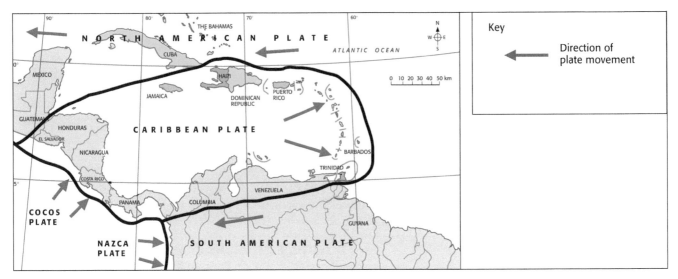

▲ *Figure 4.5 Southern Caribbean plate boundary*

Exercise

The Caribbean plate has two convergent margins, two transform margins and one very small possible area of divergence, as shown in Figure 4.5.

1 Name two Caribbean countries that lie on each type of boundary.

2 List the Caribbean countries that are not on the Caribbean plate. State which plate each one lies on.

	Convergent margin	**Divergent margin**	**Transform margin**
Plate material	Destroyed	Created	No change
Plate movement	Towards each other	Away from each other	Sliding past each other
Relation to mantle	Downwards into mantle	Upwards from mantle	No vertical movement
Landforms produced	Fold mountains, volcanoes, trenches	Mid-oceanic ridges, rift valleys, volcanoes	Major faults

▲ *Table 4.1 Differences between plate margins*

Note: All plate margins are associated with earthquakes. The movement of these large crustal slabs causes the Earth to shake.

Go to www.volcano.und.nodak.edu/vwdocs/vwlesson/plate_tectonics/part10.htm for an introduction to plate tectonics.
Go to www.scotese.com/caribanim.htm to view an animation of the evolution of the Caribbean Sea.

Formation and distribution of earthquakes, fold mountains and volcanoes

Earthquakes

The formation and distribution of **earthquakes** are closely related to plate margin activities. Whenever plates move, the Earth shakes.

✳ **Earthquakes** are vibrations or tremors in the Earth. They may be caused by the Earth's natural movements or by human explosions.

Earthquakes can be felt through the whole Earth. Knowledge of the inside of the Earth is gained from the transmission of earthquakes. The place where the movement originates has the strongest vibration and is called the **focus**. Earthquakes often start deep in the Earth's crust but they may also start on the surface. The point on the surface above the focus is called the **epicentre**. The vibration is weaker with increasing distance from the epicentre.

➤ *Figure 4.6 Anatomy of a major earthquake*

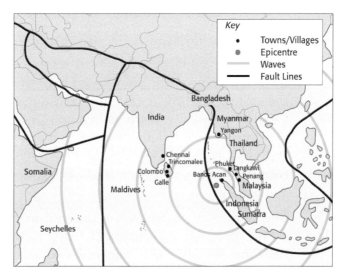

Earthquake **tremors** can vary in their intensity. During a strong earthquake the ground may open to form a **fissure** or large crack in the ground. Other strong earthquakes which occur underwater can generate very large, fast-moving sea waves called **tsunamis**. All these features created by earthquakes have a severe impact on human activities and are a major natural hazard (see Chapter 14).

Earthquakes occur frequently at plate margins. Movements of these large crustal fragments cause the Earth to shake. This is especially true at convergent margins where magma is escaping through the crust as one plate melts back into the mantle. Volcanic activity produces earthquakes.

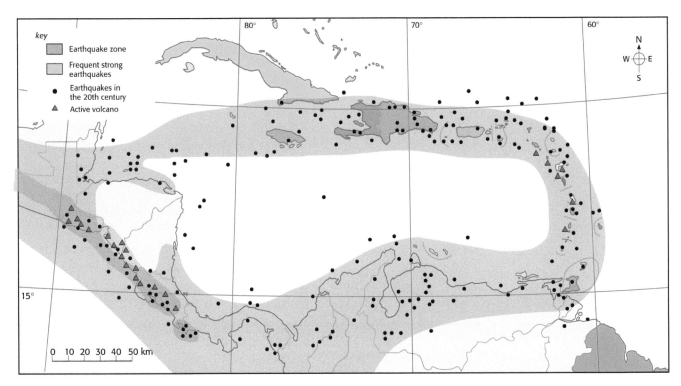

▲ *Figure 4.7 Distribution of earthquakes and volcanoes at Caribbean plate margins*

Some places far away from plate margins experience few earthquakes and are said to be stable areas.

In the Caribbean, the convergent margins of the Caribbean plate in the eastern Caribbean (Dominica, Montserrat and Martinique) and Central America, and transform margins to the north (Jamaica, Haiti) and south (Trinidad and Tobago), all experience frequent earthquakes. Barbados is on a stable area and has few earthquakes. The mainland country of Guyana lies on the South American plate and is also a very stable area.

The University of the West Indies Seismic Research Unit in Trinidad closely monitors earthquake activity in the Caribbean. Very sensitive instruments called **seismographs** continually record the tremors of the Earth in the region. Seismographs are also placed around the active volcanoes, such as those in Montserrat and Dominica, to monitor their activity.

Look for current information on Caribbean seismic activity in the region at: www.seismicunit.org

Fold mountains

Fold mountains occur at convergent plate margins.

❋ **Fold mountains** are layers of rock which have been pushed together and bent upwards to form mountains.

Fold mountains occur where plates come together at convergent margins. When the plate carrying continental materials goes back into the mantle, the lighter old rocks are scraped up and folded on the surface. They are too light to be subducted into the dense mantle. The lighter continental material 'floats' on top and becomes compressed in large folds. This happens when the rocks are relatively soft (sediments). These large folds form high mountain areas. There are many fold mountains in the world which are still rising. The highest range is the Himalayas, in Asia, which rise to 8 848 m at Mount Everest. This is the highest point above sea level on the Earth's surface. Both the Indian and Asian plates carry soft continental sediments. (Marine fossils can be seen in rocks very high above sea level.)

➤ *Figure 4.8 Formation of the Himalayas by colliding plates*

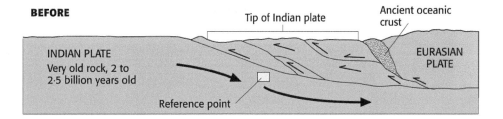

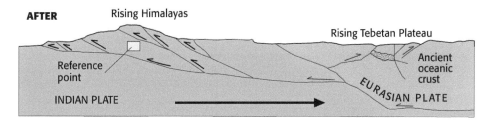

Folding in the Caribbean has resulted in two main lines of fold mountains created from the convergence of plates millions of years ago:

- the north belt of the Greater Antilles-Blue Mountains, Jamaica (2 728 m), to the Cordillera Central of Hispaniola

- in the south the Northern Range, Trinidad, which rises to 940 m.

Exercise

Using an atlas, draw and label a world map to show one fold mountain range on each continent: Himalayas, Rockies, Andes, Alps, Atlas Mountains.

➤ *Figure 4.9 Folded rocks near Gordon Town, Jamaica*

Volcanic landforms

Volcanic landforms are among the most pronounced and dramatic features associated with plate margins. Plate movements allow magma (hot molten material in the mantle) to move up into the crust as intrusive landforms or escape onto the Earth's surface forming extrusive landforms.

✳ **Extrusive volcanic landforms** are formed when magma (which is called **lava** when it is on the surface) is poured out onto the surface of the Earth. It forms volcanoes and lava plateaus.

✳ **Intrusive volcanic landforms** are created by magma cooling inside the crust. This intruded magma forms sills, dykes and batholiths in surrounding rocks.

Extrusive landforms

Cone or crater

Secondary vent and cone

Ash layers

Rocks of crust

Pipe or vent

⬒ *Figure 4.10 Structure of a composite volcano*

Volcanoes

✳ **Volcanoes** are extrusive volcanic landforms, and are formed by the Earth's magma erupting onto the surface (lava) and cooling.

Volcanoes can be divided into three main types: composite cones, acid lava cones and basic lava cones. Each kind of lava produces a different shape of volcanic cone and type of eruption.

1 Composite cones

Some volcanic cones are composed of alternating layers of lava and ash. These are called composite cones.

2 Acid lava cones

The lava is referred to as 'acid' because it is high in silica. It is very viscous (thick) and slow-moving, and cools relatively quickly. This type of lava tends to block its own vent (opening) until pressure causes it to erupt. It erupts explosively, expelling clouds of ash, gases and lava. These eruptions create steep-sided volcanoes topped by craters (a steep-sided depression at the top of the volcanic cone). Some volcanoes are so explosive that they blow away the whole of the cone and subside into a very large depression called a caldera.

➤ *Figure 4.11 Structure of Caribbean volcanoes – a caldera*

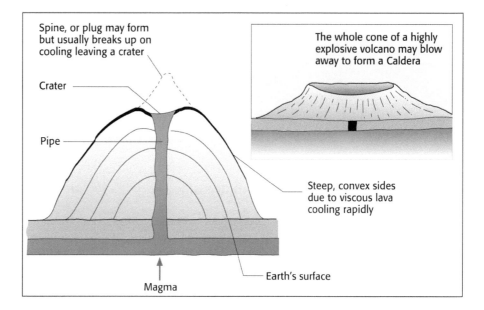

The volcanoes of the Caribbean are of this explosive type. There are many active volcanoes in the Caribbean including Langs Soufrière, Montserrat, and the submarine Kick 'em Jenny off northern Grenada. Islands of the eastern Caribbean, from St Christopher – Nevis in the north to Grenada in the south, are all volcanic; Guadeloupe is half volcanic and half limestone; while Antigua is all limestone so has no volcanic activity (see Figure 4.7).

▲ *Figure 4.12a Soufrier Hills (before eruption)*

▲ *Figure 4.12b Soufrier Hills (after eruption)*

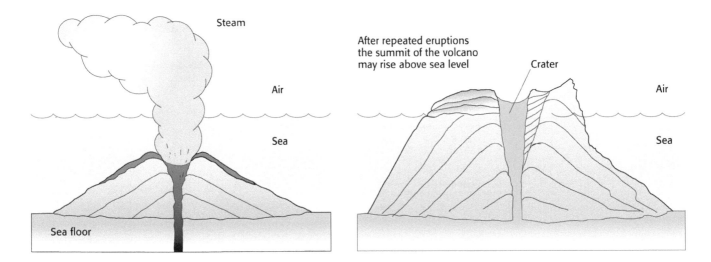

▲ *Figure 4.13 Submarine volcanoes*

3 Basic lava cones

Basic lava has less silica, and is hotter and more fluid, than acid lavas. Basic lavas cool slowly and flow over long distances. Basic lava cones may erupt continuously without explosion. This lava forms gentle-sided volcanoes composed of successive lava flows. The Hawaiian volcanoes are formed from basic lava.

▲ *Figure 4.14 Eruption of basic lava in Hawaii*

Lava plateaus

Basic lava which flows out from a long vent or fissure may form a flat, lava plateau. This feature does not occur in the Caribbean but one example may be found in western India.

Intrusive landforms

Magma which cools inside the crust may form a variety of shapes.

- **Batholiths** are the largest and deepest intrusive features. They are often the magma chambers of volcanoes.

- **Sills** are smaller intrusions along the bedding layer and form horizontal layers of magma. One example is the Newcastle Sill, Jamaica.

- **Dykes** are diagonal or vertical intrusions across the bedding planes. They form vertical walls of magma.

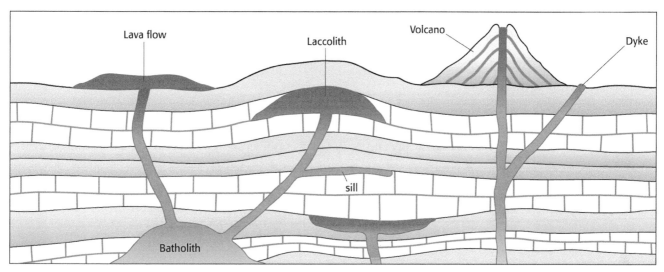

▲ *Figure 4.15 Intrusive volcanic landforms*

Changes in volcanic landforms over time

Change in eruptions

Over time, volcanoes may become dormant (do not erupt for a hundred years) or may become extinct (not known to have erupted in human history). The longer they remain dormant, the more they show the effects of weathering, mass movements and agents of erosion.

However, previously extinct volcanoes may suddenly become active. This was what happened to Krakatoa. It was a heavily populated agricultural island near Java, with fertile, volcanic soil. In August 1883, the volcano exploded and blew away two-thirds of the island.

The effects of weathering and erosion

Weathering and erosion may wear away volcanic cones and the Earth's surface to reveal intrusive landforms. Some volcanoes in the southern area of the eastern Caribbean, such as St Lucia and Grenada, show evidence of older landforms. Their hillsides have been subject to landslides, and river action has created steep-sided valleys. In St Lucia, the ash and lava have been removed from the cone leaving only the resistant volcanic pipes known as the Pitons. The Soufrière basin, including the bay, is a large caldera which has been drowned on one side and eroded by the Soufrière River.

There are also many sulphur hot springs in St Lucia. These occur when water in the rocks is heated by the magma and escapes to the surface as steam and hot water.

➤ *Figure 4.16 Sulphur springs*

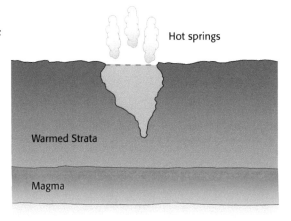

Hot springs

Warmed Strata

Magma

CHAPTER 5

External forces and limestone environments

River picking up and transporting materials

Weathered fragments

Rock

Sediments in suspension

By the end of the chapter students should be able to:

- define **denudation**, **weathering**, **mass wasting** and **erosion**

- explain the **processes of weathering**

- describe **landslides** and **soil creep**

- describe **conditions influencing the occurrence of landslides** and **soil creep**

- describe the **characteristics of limestone**

- explain the **processes operating in limestone landscapes**

- explain the formation of **karst landforms in the Caribbean**.

When mountains and volcanoes are created by processes associated with plate tectonics (see Chapter 4), they are exposed to the effects of atmospheric conditions and processes acting on the surface of the Earth. The climatic elements, rivers, ice and the waves of the sea, all act to reshape the newly created landforms. Highlands are lowered and new land created in the sea in an endless cycle of **uplift** and **denudation**.

✳ **Denudation** is the name given to all the processes which strip away the surface of the Earth, revealing the underlying rocks. It includes all the processes of weathering, mass wasting and erosion.

✳ **Weathering** refers to the break-up (disintegration) and decay (decomposition) of rocks *in situ* (which means *where they lie* on the Earth's surface), by atmospheric conditions. Weathered materials are *not* moved or transported by weathering processes. They are moved downslope by gravity as mass wasting or are removed by water.

✳ **Mass wasting** is the movement of weathered materials (with or without water or ice) downslope by the force of gravity.

✳ **Erosion** refers to the processes which **break up**, **transport** and **deposit** weathered materials and rocks. The agents of erosion are rivers, waves, ice and wind.

Weathering processes

The processes of weathering are facilitated by two important elements of climate: temperature and rainfall.

The processes may be divided into three types – physical weathering, chemical weathering and biotic weathering – according to the effect they have on the rocks. The result of the processes depends on the characteristics of the rock. The chemical composition of the rock, its structure and formation will influence how it responds to weathering by atmospheric conditions and plants.

✳ **Physical/mechanical weathering** results in the break-up of rock into small pieces with no change in their chemical structure. The rock **disintegrates**.

✳ Processes of **chemical weathering** change the chemical structure and physical appearance of the rocks. The rock **decomposes**.

✳ **Biotic weathering** is the result of the physical or chemical effects of plants and animals on rocks.

Physical weathering processes

These processes tend to be most active in cold climates and dry climates. Pressure release can occur under any conditions.

● **Pressure release:** when rocks that are formed under great pressure (volcanic rocks, for example) are exposed on the surface, the release of the pressure of overlying layers causes them to expand, crack and break up. Pressure release forms areas of large broken blocks of rock.

➤ *Figure 5.1 Pressure release breaks up volcanic rocks*

- **Frost action:** in cold climates and high mountains where there is moisture and where temperatures are around freezing, the rocks are subject to alternating periods of warmth and cold.
  - During the warm period, water seeps into pores and cracks in the rock.
  - When temperatures fall, the water in the rock freezes.
  - Ice has a greater volume than water so it expands by 9 per cent.
  - Cracks and pore spaces are enlarged by the ice.
  - When temperatures rise, the ice thaws.
  - The water seeps further into the rock.
  - It freezes again when it is cold (below 0°C).
  - Repeated freezing and thawing eventually breaks the rock into small jagged fragments. These form a pile of material called **scree** at the foot of mountain slopes. This process occurs in the highest mountains in the Caribbean on very cool nights in December.

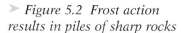

➤ *Figure 5.2 Frost action results in piles of sharp rocks*

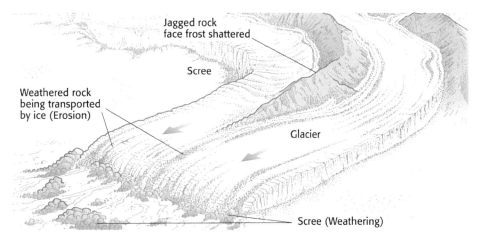

- **Temperature changes**: rocks expand and contract in response to changes in temperature. In deserts, there is a large daily range of temperatures. During the day it is very hot and the rocks expand; during the night it gets quite cool and the rocks contract. Some rocks peel off in layers in response to the repeated expansion and contraction. This is called exfoliation.

The weathered fragments stay just where they are broken up. They may be moved later by gravity as mass wasting, or be transported by agents of erosion.

Chemical weathering processes

Chemical weathering processes occur rapidly in hot, wet climates such as those in the Caribbean. They involve a chemical change in the rock. Limestone, which is made of calcium carbonate, is especially vulnerable to the chemical weathering process known as **carbonation**.

There are many areas of limestone in the Caribbean which show the effects of this process.

- **Carbonation:**
  - Rainwater absorbs carbon dioxide in the atmosphere to produce a weak carbonic acid.
  - This acid acts on the calcium carbonate in limestone to form calcium bicarbonate.
  - Calcium bicarbonate is dissolved by water.

This results in the pits and holes common in limestone areas and known as karst landforms. Residues of iron and aluminium deposits may be left over from this process, forming iron ore and bauxite ores, as in Jamaica.

➤ *Figure 5.3 Limestone being chemically weathered*

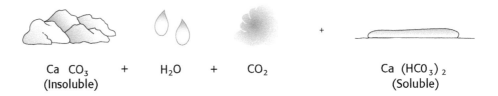

Ca CO$_3$ +   H$_2$O   +   CO$_2$ +   Ca (HCO$_3$)$_2$
(Insoluble)   (Soluble)

- **Oxidation:** this is the process of oxygen being added to a compound. Rocks rich in iron react with oxygen in water to form iron oxide, as in the rusting of iron. The effect is a thin soft crust on the outer part of the rock that is easily broken down

➤ *Figure 5.4 Oxidation*

$$4Fe + 3O_2 \text{ (in presence of water)} \Longrightarrow 2Fe_2O_3 \text{ (hydrated)}$$

- **Solution:** this occurs in rocks containing salts, such as sodium and potassium, that are soluble in water. Rainwater falling on these rocks will eventually dissolve them.

Exercise

Weathering generally occurs quite slowly. You will have to observe and measure your samples carefully in order to demonstrate that weathering has taken place.

1 Take 3–4 rock fragments, ideally about 5–10 cm long. Try to find different types: some dark-coloured, some light, hard and soft rocks.

2 Put a sheet of plastic in an open box in an area of the school that is exposed to rain and sun.

3 Measure the rock fragments and put them on the plastic sheet.

4 Leave them out in the open for 3 months (start to end of term, for example).

5 Then measure them again to see if they are smaller.

6 Collect any rock fragments around the rocks.

7 Can you identify which weathering processes may have been at work on the rocks?

Biotic weathering

Biotic weathering is recognised as a distinct type of weathering, since plants and animals living on rock surfaces can cause the disintegration or decomposition of the rocks. As they grow and die they change the rocks.

- **Plant roots** get into cracks and spaces of a rock and break it apart as they grow. This is a physical process, leading to disintegration of rocks.

- **Decayed vegetation** may release organic acids which chemically alter rocks.

Mass wasting

After rocks are weathered (disintegrated or decomposed) they accumulate to form the beginning of soils to which decayed vegetative matter (humus) will be added. If the weathered material is on a slope, even a very slight incline, the force of gravity will move it downwards.

The slowest of these movements is **soil creep**. We cannot see this movement but we can observe its results.

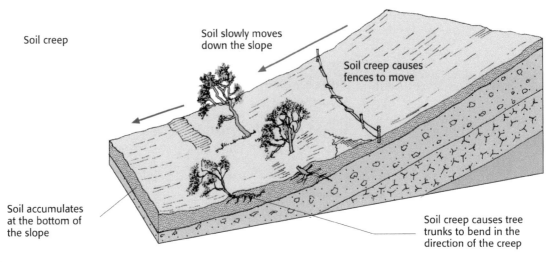

Soil creep

Soil slowly moves down the slope

Soil creep causes fences to move

Soil accumulates at the bottom of the slope

Soil creep causes tree trunks to bend in the direction of the creep

▲ *Figure 5.5 Soil creep*

✳ **Soil creep** is the slow downward movement of weathered material on slopes by gravity.

Exercise

Soil creep is very slow and can only be seen by its results over time.

1 Find a piece of stick or a tree branch about 2 m long and strong enough to stick in the ground.

2 Place it firmly into the soil vertically, making sure it is at least 0.5 m below the surface.

3 Observe the stick after 6 months and see if there is any change in its position; it may be leaning outwards as a result of soil creep, especially if it is on a slope.

➤ *Figure 5.6 Hazardous landslides block a road*

✳ **Landslides** are faster movements of weathered materials, often with water, down steep slopes under the force of gravity.

Conditions influencing landslides and soil creep

- Nature of the rock: areas made of clay are prone to landslides because the clay is very slippery when it is wet and causes the overlying soil/weathered material to slide.

- Size of individual grains of weathered material: small particles will be more stable than larger particles and therefore remain at rest even on steep slopes.

- Amount/weight of weathered material on slope: this will determine the stability of the whole slope.

- Angle of slope and resolution of force of gravity on slope: gravity will have a stronger effect on very steep slopes than on gentle slopes. This accounts for the very slow movement of soil on gentle slopes.

- Amount of rainfall causing slide of the weathered materials: water adds weight to the weathered material and also acts as a lubricant, making it easier for the material to move. Heavy rainfall often triggers landslides in the Caribbean.

In the Caribbean, the processes of weathering and mass wasting are very active. Landforms produced by volcanoes and folding are lowered and reshaped by these processes. The hot, wet, tropical marine climate of the Caribbean (see Chapter 8) results in rapid and deep weathering of exposed rocks. Weathering is responsible for one of our natural resources – bauxite in Jamaica and Guyana (see Chapter 12).

Mass wasting is especially active on the steep volcanic and fold mountain slopes in the Caribbean during torrential rainfall. (Limestone areas do not have many surface rivers because the water goes underground. There is also little mass wasting because the weathered materials are dissolved.)

Rocks respond to weathering and mass wasting in different ways. Rocks with layers may break down in layers, whereas those with large grains may break down into grains. Similarly the chemical composition of the rock will influence how it is weathered chemically. The Caribbean has many different kinds of rocks: igneous rocks formed by volcanoes, sedimentary rocks such as limestone and sandstone, and metamorphic rocks, like marble, which are changed by heat or pressure.

Limestone environments

Many areas of the world are made up of accumulations of the calcium carbonate remains of marine organisms.

Belize
The geology of Belize consists of different varieties of limestone. The hilly regions surrounding the (non-limestone) Maya Mountains exhibit typical karst landforms such as sinkholes, caverns and underground streams.

Barbados
Barbados is a coral capped island — coral limestone overlays sedimentary rock. The main limestone areas are in the northeast of the island but there is a lower limestone ridge in the Christ Church area.

Jamaica
Nearly two thirds of Jamaica is covered by White Limestone and there are also large areas of Yellow Limestone. Highly-developed *karst* landscape is found in the Cockpit country. Jamaica has the highest number of caves per kilometre in the world.

▲ *Figure 5.7 Some limestone environments of the Caribbean*

Characteristics of limestone

- Chemical composition: more than 80 per cent calcium carbonate
- Structure:
  - distinct joints/cracks
  - layers along bedding planes
  - coarse texture with large pores
- Permeability
  - very permeable, i.e. water passes through limestone very easily; the water also goes down the holes created by chemical weathering (carbonation)
- Light-coloured
- Relatively soft but varies with age.

Processes in limestone areas

There are two basic processes operating in areas of limestone.

- **Carbonation**: limestone is changed and dissolved away, forming pits and holes. This process may continue underground until the roof of the underground feature collapses to form depressions and steep valleys (passages).

- **Evaporation/deposition:** a thin layer of calcium bicarbonate in solution is deposited in caves as the water drips off the ceiling onto the floor.

These processes result in water flowing underground, leaving dry surface features and varied underground features which are referred to collectively as **karst landforms**.

Karst landforms of the Caribbean

❋ **Karst** is the name given to all the landforms in areas of limestone rock. It includes both surface features and underground features.

Karst landscapes

There are three types of karst landscape: tropical karst, cockpit karst and temperate karst. The most common types found in the Caribbean are tropical karst and the distinct cockpit karst in Jamaica.

Limestone rock is composed mainly of calcium carbonate. In the Caribbean there were long geological periods when shallow water formed coral reefs which were then raised above sea level to form land. The older Caribbean limestones are the yellow and white groups of the Greater Antilles. These are millions of years old. The limestone of the eastern Caribbean is more recent, being less than one million years old.

Limestone rock is also a valuable resource. It is quarried and used for building houses and for making cement.

Guadeloupe, one of the French islands of the eastern Caribbean, is unique in being composed of both volcanic material and limestone layers. It marks the southern end of the outer limestone-covered arc of islands which includes Antigua and Anguilla.

➤ *Figure 5.8 How the type of rock influences landforms in Guadeloupe*

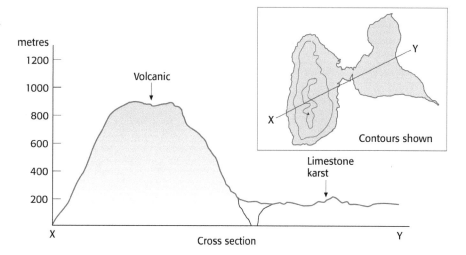

Karst landforms can be divided into those that are formed on the surface and those formed underground. Underground features may eventually be exposed on the surface as the overlying layers collapse.

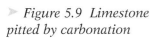
Table 5.1 Surface and underground limestone landforms

Surface	Underground
Dry valleys, gorges	Caves, passages
Swallow holes	Stalactites, stalagmites, pillars
Depressions, cockpits	
Limestone pavements	

Surface features

Limestone pavements – clints and grykes
When rain falls on limestone, carbonation in weaker joints and cracks causes pitting and wearing of the surface into grooves, leaving small ridges in between. These ridges are **clints** and the worn areas are called **grykes**. In the Caribbean, exposed limestone is often pitted and grooved by this process. Clints and grykes can be seen on the St Lucy tableland of northern Barbados.

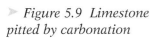
Figure 5.9 Limestone pitted by carbonation

Dry valleys
During the rainy season water may collect and flow over the surface as temporary streams in steep-sided valleys. For most of the year, though, they are dry. There are many of these **dry valleys** along the west coast of Barbados, where they are called 'gullies'. Dense

vegetation often masks their course as plants take advantage of water near the surface. In the Central Uplands of Barbados, Welchman Hall Gully and Jack-in-the-Box Gully appear to be collapsed underground passages. Their steep sides, the presence of stalactites, and their connection to existing underground caves and passages, support this suggestion.

Swallow holes

Carbonation often works to enlarge joints in the limestone to form **swallow holes**. These holes then allow rivers to pass underground. Sometimes they are large enough to form steep-sided depressions. Some depressions may have debris at the bottom and so small ponds are formed. (It has been suggested that the flat bottoms are the result of periodic flooding of the depressions.)

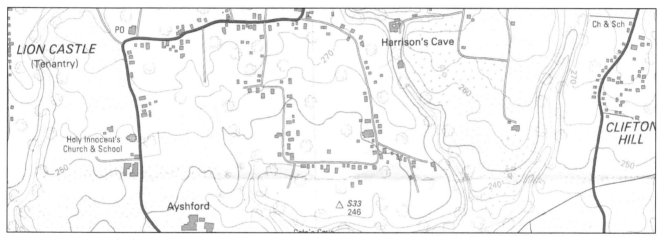

▲ *Figure 5.10 A relief map showing tropical karst formations in central Barbados*

Underground features

Most of the water infiltrates (seeps) into the surface of the limestone through the pores and cracks. If the rainfall is intense, some water may flow over the surface, forming a temporary/seasonal stream. Eventually the surface water flows through any holes formed by enlarged joints (**swallow holes**). The now underground water flows along bedding planes and joints, dissolving the rock to form large underground **passages**, **caves** and **caverns**.

As the water drips from ceiling to floor, evaporation causes deposition of thin layers of calcium out of solution.Build-up of calcium forms **stalactites**, **stalagmites** and **pillars**.

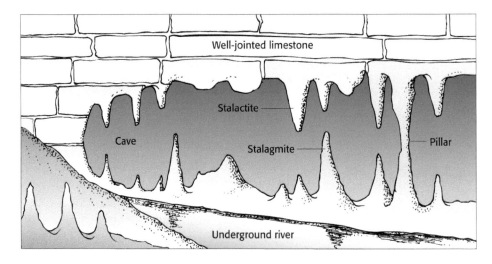

Well-jointed limestone

Stalactite

Cave

Stalagmite

Pillar

Underground river

▲ *Figure 5.11 Interior view of a limestone cave*

Exercise

1 Build a model of limestone surface and underground features using clay or *papier-mâché* to mould dry valleys and underground caves with associated features.

2 Try pouring water on the surface and watch it flow through the swallow holes and go underground.

You can see a film of this process at www.cockpitcountry.com/landscapeimgs/film.cc.swf

Cockpit karst in Jamaica

White limestone covers more than 70 per cent of the area of Jamaica. In central Jamaica, joints and faults at right-angles have given rise to a particular type of karst named for this area: cockpit karst. It consists of a distinct hummocky 'basket-of-eggs' topography. Small hills of equal height are interspersed with depressions. These are formed when carbonation is rapid at the intersection of the joints, widening them into depressions and leaving the surrounding areas as hills.

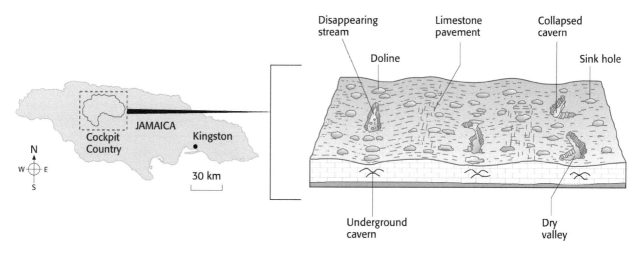

Figure 5.12 Cockpit karst in Jamaica

You can view excellent diagrams and a discussion of the formation of cockpits (depressions) and conical hills at www.cockpitcountry.com/Formation

---

 ✳ A **cockpit** is a steep-sided depression found in limestone.

 ✳ **Conical hills** are the remnants of the original limestone surface, usually all at the same height above sea level.

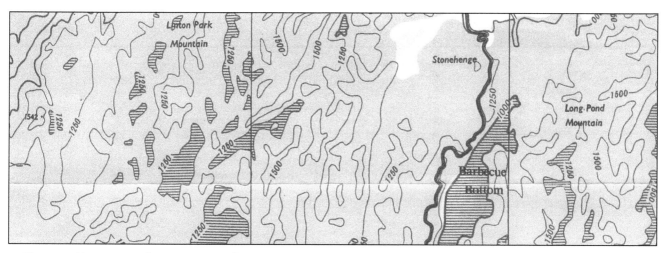

Figure 5.13 A map of Jamaican Cockpit Country

Exercise

1 Name one large depression on Figure 5.13.

2 Describe the drainage on the map (note springs and ponds).

3 What is the height of the highest points?

4 Can you identify the height of the original surface?

CHAPTER 6

Rivers

By the end of this chapter students should be able to:

- describe the **water cycle**
- describe how **water flows when it reaches the Earth's surface**
- describe **river processes**
- explain the **formation of river valleys and river channels**
- describe **trellis, radial and dendritic drainage patterns**.

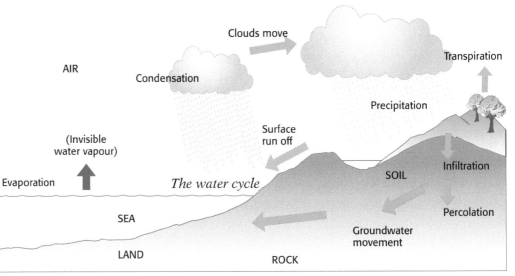

The water cycle

▲ *Figure 6.1 The water cycle*

The water cycle

The transfers of water between the air, sea and land are considered a cycle because water changes in form but not in amount, as it moves between these locations. There are several different transfers between the land, sea and air.

- **Precipitation:** this includes all forms of moisture falling from the air on both land and sea under gravity – rain, ice, snow, sleet (see Chapter 8).

- **Surface run-off:** this includes all water running from the land into the sea under gravity – rivers, underground water, unchannelled water, glaciers.

- **Evaporation:** the process when moisture changes from a liquid in oceans, lakes and rivers to a gas in the air (water vapour).

- **Condensation:** the process when moisture changes from a gas into a liquid in the atmosphere as clouds.

- **Transpiration:** the process when moisture loss from plants transfers water from land to air.

See website www.physicalgeography.net/fundamentals/8b.html for further study of the water cycle.

Pathways water takes on reaching the Earth's surface

On reaching the Earth's surface, some water remains on the surface and some moves through the soil and the rocks towards the sea.

- **Surface run-off** may be divided into water in channels, i.e. rivers, and unchannelled water such as sheet wash, rills and gullies.

- The remainder of precipitation goes into the soil. This is called **infiltration**.

Infiltration

✳ **Infiltration** is the entry of water into the soil. Water soaks into the ground.

Some soils have many spaces between the particles and allow a lot of water to sink in or infiltrate (sandy soils). Other soils have very few small spaces which are quickly filled and do not allow any more water to infiltrate (clay soils). Vegetated slopes generally have more infiltration than bare slopes. The vegetation intercepts the rainfall, so it drips off the leaves and flows down the stem, reaching the ground surface slowly and in small amounts.

✳ **Throughflow** is the downward and lateral movement of water through the soil under gravity.

✳ **Percolation** is the continued downward movement of water through the spaces in the rock.

Percolation occurs until all the spaces in the rock are filled with water. Some rocks, like limestone, have a lot of spaces and water easily percolates through them; they are **permeable** (see Chapter 5). Other rocks have few spaces and will not let water pass through them; these rocks are **impermeable**.

✳ The **water table** is the level below which the rock spaces are filled with water. It may be referred to as the **level of saturation** in the rock. This level may vary according to the amount of precipitation and water use.

Groundwater eventually enters the sea directly as springs along the coastline or by springs feeding rivers.

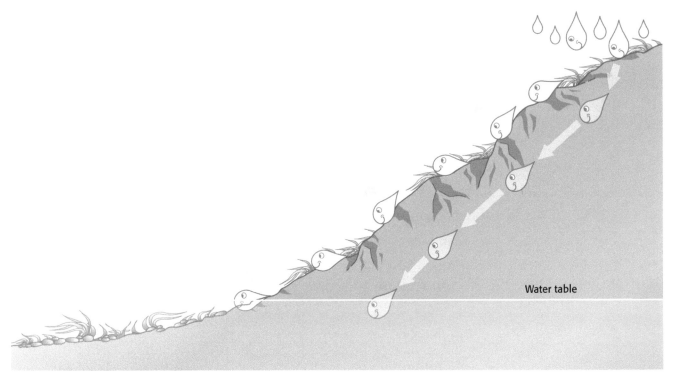

▲ *Figure 6.1 Water moving to the sea*

Go to www.physicalgeography.net/fundamentals/8m.html for diagrams on groundwater movements.

Rivers

In Chapter 5 you learned that rivers are one of the agents of erosion that pick up, transport and deposit weathered materials and rocks. River action is the major force wearing down mountains and depositing silt into the sea. These river deposits may in turn be uplifted to form new land as folded and faulted sediments.

Rivers occur on all continents and act on most of the world's landscapes. Some even occur in deserts, having their source outside the region, e.g. the Nile flowing through the Sahara Desert, or the Colorado River flowing through the deserts of Arizona, USA. In cold areas, melting ice is the source of large rivers such as the Rhône River in France. The longest rivers, which drain large areas of the continents, are the Nile, Mississippi–Missouri, Amazon and Ganges.

➤ *Table 6.1 The world's longest rivers*

Name	Location	Length in km (miles)
Nile	Africa	6 690 (4 160)
Amazon	South America	6 404 (4 000)
Yangtze	Asia	6 378 (3 964)
Mississippi–Missouri	North America	6 021 (3 710)
Ob	Europe	5 567 (3 459)

Go to www.rivernile.com for more information on the River Nile.

Caribbean rivers

The longest rivers in the Caribbean are found on the mainland territory of Guyana. The Essequibo and Demarara rivers flow from the wet equatorial areas to their mouths in the Atlantic Ocean. The volcanic Caribbean islands are geologically young, so the rivers are generally straight along their courses. On the larger islands, some rivers have developed large valleys such as the Rio Grande and Wagwater rivers of Jamaica, and the Caroni River of Trinidad.

➤ *Figure 6.2 A comparison of the Essequibo River and the rivers on some Caribbean islands*

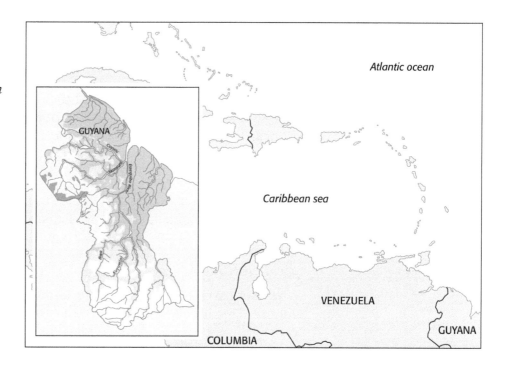

Go to www.pr.water.usgs.gov/public/caribbean/index.html to find out more about rivers in the Caribbean.

River processes

As the river flows from its source to its mouth it changes the land by **erosion**, **transportation** and **deposition**. After the river has carved out the valley, weathering processes also act on the exposed valley sides.

Erosion

Rivers pick up and wear away materials by different processes of erosion.

- **Corrasion** is the removal of soil and rocks by the use of other materials. The formation of pot-holes is one result of the process of corrasion. Pebbles carried by the river get caught in rough areas of the bed of the channel. They swirl around like drills, creating holes in the bedrock and wearing it down.

- **Hydraulic action** refers to the sheer force of the water in wearing away the land. This is often seen on the banks of the channel, where water flows around a bend and crashes into the bank, undercutting it and causing it to collapse. This causes widening of the channel, and eventually widening of the valley floor.

- **Corrosion** is the chemical action of water on the rocks. Some rocks, such as limestone, react chemically with water and are worn away.

- **Attrition** is the wearing way of the rocks carried by the river as they collide with each other. The river may pick up large boulders near its source but these are reduced to fine particles of silt as they approach the sea. The rocks collide, smoothing and wearing each other down.

These processes act to cut down the river channel as vertical erosion, or widen the channel and valley as lateral erosion.

Transportation

The river carries the eroded materials as it flows from source to mouth. This is called the **load** of the river.

⁕ The **load** of the river is all the materials carried by the water.

The load is transported in four ways:

- **Traction:** the largest boulders are *dragged* along the river bed.

- **Saltation:** smaller rocks and pebbles are *bounced* along the bed.

- **Suspension:** very fine particles (clay/silt) are *held up* by the water itself – they give the water its colour.

- **Solution:** some soluble minerals are chemically *dissolved* in the water.

▼ *Figure 6.3 Methods of transportation*

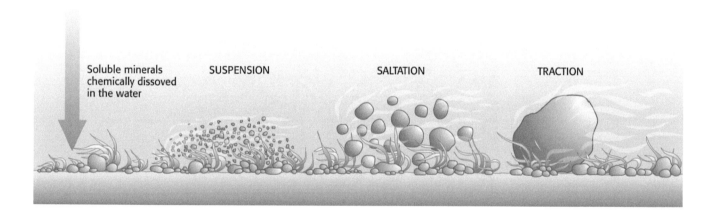

Soluble minerals chemically dissoved in the water

SUSPENSION

SALTATION

TRACTION

Exercise

Take a bottle of water from a river and let it settle for a week. At the end of that time, at the bottom of the bottle you will see the load which was in suspension.

Deposition

This process occurs when the river drops its load. It need not be a permanent feature as the river can pick up and put down materials according to the amount of water in the channel or its speed. When in flood, the river carries a lot of material, but as the water level drops, it deposits its load. It may pick up the load when it floods again. During the hurricane season, rivers in the Caribbean turn into rushing torrents carrying a large load. But in the dry season they are reduced to a trickle, having deposited their load all around, until the next hurricane season. In Figure 6.5 notice the size of the boulders compared with the small amount of water in the river.

> *Figure 6.4 Deposition in a river channel*

Go to www.physicalgeography.net/fundamentals/10y.html to learn more about stream flow.

River valleys and river channels

It is important to distinguish between a river **valley** and its **channel**. Note that the channel is **in the valley**.

❋ A **valley** is the long depression in which a river flows. It includes the channel sides and floor and the flowing water.

❋ A **channel** is the actual place where the water is flowing. It is the wet part of the valley, consisting of banks and bed.

➤ *Figure 6.5*
A river channel

Exercise

Try creating your own little river by pouring water slowly onto a pile of loose material, e.g. sand on a beach or in a waterproof tray. As the water flows you will see it making a small channel and flowing downwards by gravity.

✳ A **drainage basin** is the area drained by the river and its tributaries.

The parts of a drainage basin are named in Figure 6.6.

➤ *Figure 6.6 Parts of a river basin*

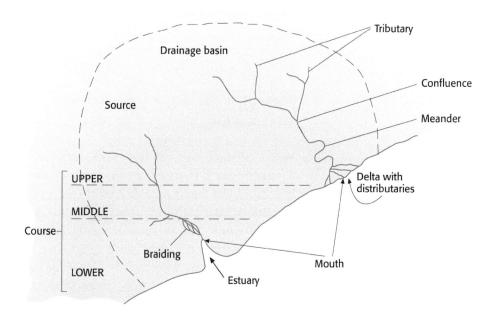

The formation of river valleys and river channels

The river processes take place along the entire course at different times, causing changes in the channel and valley. At first, erosion tends to dominate as the river strives to flow at sea level, but later on more deposition may occur as it approaches the sea.

Upper course

Waterfalls and rapids

When rivers flow down a steep slope, the water may fall over a rock edge as a **waterfall**. There are many waterfalls in the Caribbean. The most spectacular is the Kaieteur Falls of the Potaro River (a tributary of the Essequibo) of Guyana. The water falls over the edge of the old, hard Guyana Plateau, free-falling 244 m. The highest waterfall in the world, the Angel Falls, lies just across the border in Venezuela as the Cuyuni (also a tributary of the Essequibo) flows off the Plateau (979 m). There are 20 large waterfalls in Guyana alone.

There are many other smaller waterfalls in Tobago, Dominica and other islands. The unusual Dunn's River Falls in Jamaica fall straight into the sea. This suggests a recent uplift of the land.

➤ *Figure 6.7 Parts of a waterfall*

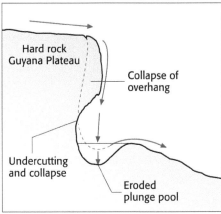

Waterfalls are formed when a river flows over a layer of hard rock. The hard rock is resistant and erodes more slowly than the soft rock below it. The hard rock forms a steep edge over which the water falls. Gradually, the falling water digs out a deep pool at its base, called a **plunge pool**. The impact of the falling water (hydraulic action) undercuts the cliff edge until it collapses. This causes the waterfall to retreat upstream. Eventually, the river may wear down the entire steep edge to leave a more gentle slope with the water flowing rapidly over it. These areas of turbulent water are called **rapids**.

Middle course

River cliffs and point bars

More water enters the channel as tributaries join it. The increase in the amount of water makes the river more powerful as it crashes into its banks. The force of the water on its banks, especially on the outside of the bend, causes erosion. This undercuts the banks, which collapse, forming a steep-sided **river cliff**. On the inside of the bend, the slack water may result in deposition. This deposited material is called a **point bar**.

▼ Figure 6.8 River cliffs and point bars

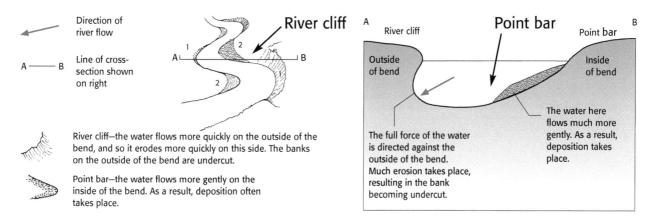

Direction of river flow

A ——— B   Line of cross-section shown on right

River cliff—the water flows more quickly on the outside of the bend, and so it erodes more quickly on this side. The banks on the outside of the bend are undercut.

Point bar—the water flows more gently on the inside of the bend. As a result, deposition often takes place.

Go to www.cscc.edu/DOCS/BIO/jamaica/graphics/RIVERS.jpg for more photographs of rivers in Jamaica.

Both erosion and deposition occur in the middle section of the channel. Some rivers that periodically carry a heavy load may develop **braiding**. This is the deposition of the load in its own channel, which forces the river to flow around its own sediments in smaller channels. These rejoin into one channel when the river is again able to carry the reduced load. The small islands of material in the channel are called **bars**.

➤ Figure 6.9 Braiding on the Essequibo River, Guyana

Lower course

Meanders and ox-bow lakes

The river continues to widen its valley until the channel is no longer in contact with the valley sides. The river channel takes a bending path over the gently sloping land. These bends in the channel are called **meanders**. A meander may become cut off from the main river by erosion and deposition. Erosion is stronger on the outer bank of the bend while deposition continues on the inner bank. This continues until the river flows straight through the narrow neck of land, leaving a separated crescent-shaped lake referred to as an **ox-bow lake**. This process is shown in Figure 6.12. These lakes are not permanent and eventually dry up. In this way the river channel may itself change its course many times.

▼ *Figure 6.10 Formation of an ox-bow lake*

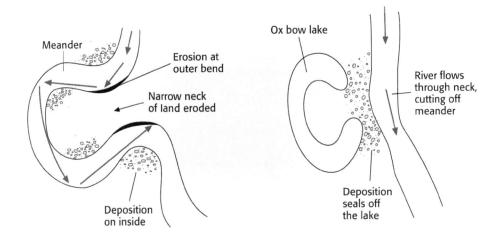

Figure 6.11 Lower course of a river valley and its meandering channel

A river in its lower course has a great volume of water. As it approaches its mouth, it flows over a low gradient very near to sea level. The river carries a heavy load which it may deposit as it periodically floods its banks. This flooding may cover the valley floor with sediment until the river is flowing over its own deposits rather than on rock.

Flood plain and levées
The river often floods its banks in the lower course as it has a lot of water flowing over a nearly flat surface. The capital of Guyana, Georgetown, is built on such a flat area and is often flooded. When a river overflows the banks of its channel, the water with its load of sediment spreads out over the valley floor. This gently sloping area of deposited material is called the **flood plain**. Repeated flooding builds up the bed of the channel until it is flowing on its own deposits above the valley floor. As the floodwater recedes, the largest materials are deposited first, closest to the channel. These deposits form raised banks called **levées**.

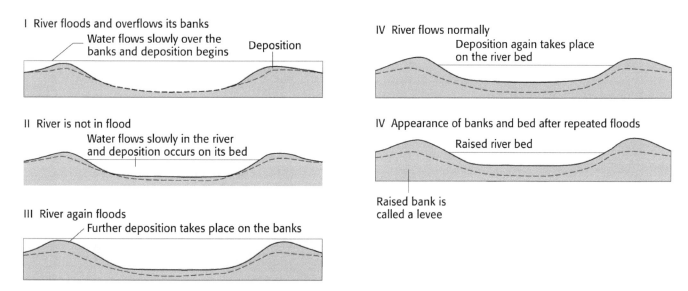

I River floods and overflows its banks
Water flows slowly over the banks and deposition begins   Deposition

II River is not in flood
Water flows slowly in the river and deposition occurs on its bed

III River again floods
Further deposition takes place on the banks

IV River flows normally
Deposition again takes place on the river bed

IV Appearance of banks and bed after repeated floods
Raised river bed

Raised bank is called a levee

▲ *Figure 6.12 Successive floods produce levées*

Floods are a natural event along a river's course. They occur when there is too much water to be contained in the channel. This can happen along any part of the course, but frequently occurs in the lower course where:

- all the water from many tributaries has joined the river

- the gradient of the valley floor is very low (almost level)

- the river is close to its mouth as it enters the sea.

In periods of heavy rainfall, or in cold countries when a river is carrying extra meltwater in the drainage basin, the large volume of water reaching the channel in its lower course cannot be contained within its banks and so it overflows. However, flooding is made worse by human activity in the drainage basin. People may:

- clear the vegetation, reducing levels of absorption and allowing greater infiltration so that there is increased run-off

- build urban structures with impermeable surfaces, which also increase the amount and rate of run-off.

Deltas

A delta is a triangular-shaped area of deposits at the mouth of a river. As it enters the sea, the river can no longer carry its load, which is deposited in the sea. The river is forced to divide into smaller channels called **distributaries**. This low-lying swampy area gradually dries out and extends the land out into the sea. These features can be seen near New Orleans, which at one time was at the mouth of the Mississippi River on the Gulf of Mexico but is now more than 100 km inland as the delta continues to build outwards.

Conditions necessary for the formation of deltas:

- The river must have a heavy load of material.

- Wave action and currents must be weaker than river deposition.

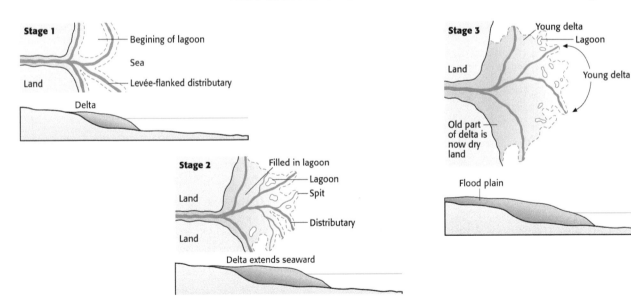

▲ *Figure 6.13 Formation of a delta*

Go to www.gsc.nrcan.gc.ca/geoscape/vancouver/fraser.asp to see an animation of the formation of the Fraser River delta in Canada.

Valleys

All the processes and landforms created by a river are found in its **valley**. Once the river has created its valley, this is further shaped by weathering and mass wasting processes (see Chapter 5). The shape of the valley is influenced by both sets of processes. The river widens the valley by lateral erosion and covers the valley floor with a smooth layer of silt.

However, some rivers flow in valleys that have been created by other processes, for example the Diego Martin River in Trinidad flows in a faulted valley. This valley is much higher and wider than it would have been if it had been formed by river processes alone.

On a map of any river course, you can identify the shape of the valley by the contour patterns and by the river channel (coloured in blue). Any sediments are shown in brown (Figure 6.16).

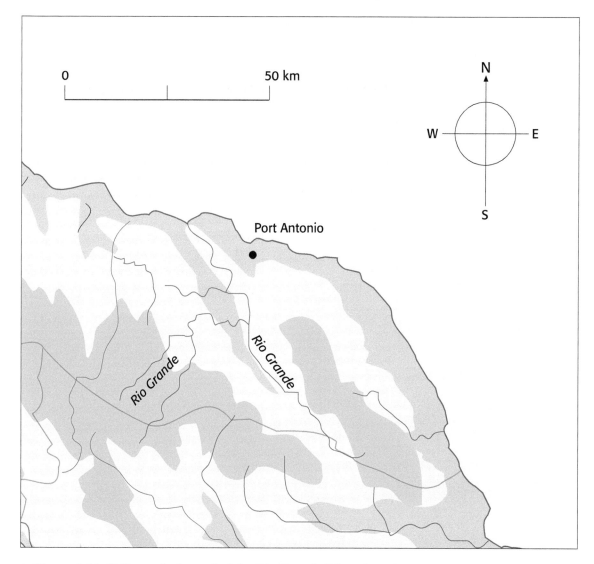

▲ *Figure 6.14 Valley and channel of the Rio Grande River, Jamaica*

Exercise

Draw a sketch map of Figure 6.14. Follow these steps:

1 Draw a frame.

2 Draw a grid similar to the one on the map (it can be smaller to fit on your paper).

   a Use a blue crayon to put in the **channel** of the Rio Grande. (*Hint*: Copy it square by square from the grid on the map onto the grid on your paper – see Chapter 2).
   b Use brown to show the deposits in the channel and to shade the flood plain.
   c Use the cliff symbol to show river cliffs.
   d Outline the **drainage basin** using another colour. (*Hint*: Follow the contour lines.)

➤ *Table 6.2 Landforms associated with sections of the valley*

Section/shape	Channel	Processes	Landforms
Upper – V-shaped	shallow, narrow	erosion	waterfalls, rapids
Middle – wider floor	asymmetrical, deeper, wider	erosion on outer bend, deposition on inner bend	river cliffs, point bars
Lower – open U-shape	wide, deep	deposition, flooding, meandering	flood plain, levées, ox-bow lakes

Go to www.fed.cuhk.edu.hk/geo/AL/core/landform/index.html to find out more about river landforms.

Gorges

Gorges (sometimes called canyons) are a particular kind of very deep valley. An entire valley may be a gorge or just a section of it.

✳ A **gorge** is a narrow, deep, steep-sided valley.

Gorges are formed by:

● the retreat of waterfalls – the falling water undercuts the cliff, causing it to collapse and leaving a gorge

- powerful rivers maintaining downward erosion as the land rises
- resistant rocks which allow very little wearing away of the valley sides.

The best example of a gorge or canyon is the Grand Canyon of Arizona in the USA. The high walls of coloured rock make it one of the wonders of the world. The river's source is in the Rockies and its mouth in the Gulf of California, Mexico.

The Grand Canyon was formed over millions of years by:

- the uplift of the Colorado Plateau
- the powerful Colorado River dropping 700 m over its 446 km – it was able to maintain its vertical erosion as the land was uplifted
- the fact that it passes through a very dry area, so there is little mass wasting to wear back the valley sides
- physical weathering, which has resulted in steep slopes and piles of coloured weathered materials on the valley sides.

Go to www.nps.gov/grca for more information on the Grand Canyon.

In the Caribbean there are gorges along the Wagwater River in Jamaica. These are the result of uplift and downcutting through very resistant rocks. The retreat of the Kaiteur Falls, Guyana has also produced a gorge.

Drainage patterns

The drainage pattern of a river is how streams are arranged on the landscape. It is possible to examine a river and its tributaries and to identify different kinds of pattern.

Dendritic

The dendritic pattern is one of the most common patterns. The river and its tributaries flowing over the same rock form a tree-like pattern. The tributaries are the 'branches', with the main stream as the 'trunk'. Many Caribbean rivers show this pattern.

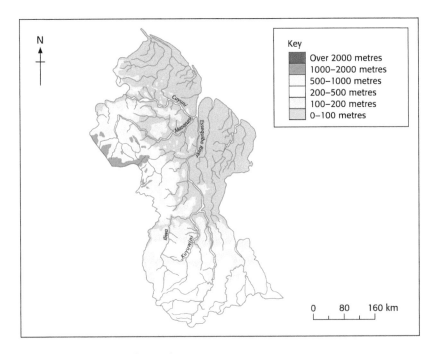

▲ *Figure 6.15 Dendritic drainage pattern*

Radial

A radial pattern is formed when all the rivers flow away from each other around a common central point. This pattern is very common in volcanic islands. The rivers have their source high on the volcanic peak and move downwards and outwards from it.

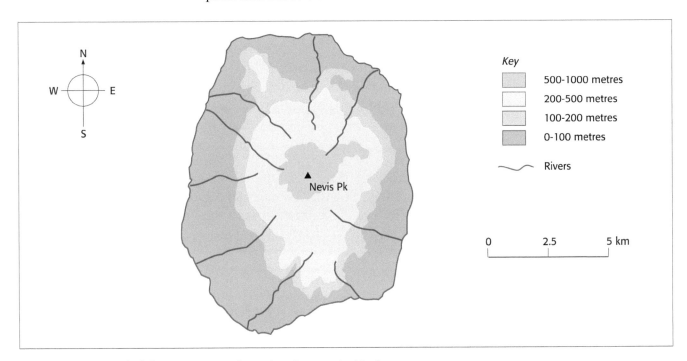

▲ *Figure 6.16 Radial drainage around a volcanic cone in Nevis*

Trellis

The trellis drainage pattern occurs under special circumstances and is the least common pattern. The river and tributaries meet at right-angles and so mimic the cross-barred pattern of a trellis. This pattern can be seen in North Trinidad.

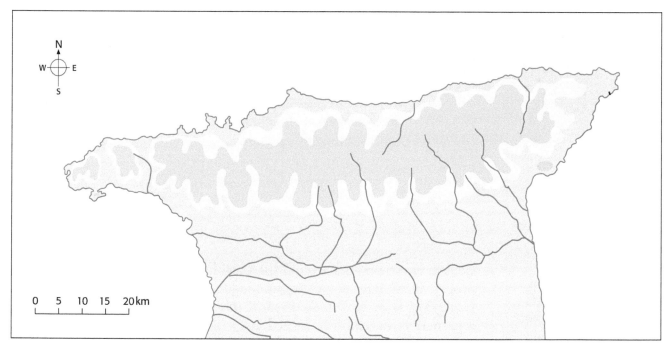

▲ *Figure 6.17 Trellis drainage pattern*

Exercise

Use an atlas to answer these questions.

1 Find a map of North America showing the Mississippi River, and one of South America showing the Amazon River.

2 Copy the course of each river and its tributaries into your book.

3 What sort of pattern do the streams make in each case?

Coasts

By the end of the chapter students should be able to:

- name and explain types of waves: **constructive and destructive waves**
- describe **wave processes**
- explain the formation of **coastal landforms**
- describe the **types and location of coral reefs found in the Caribbean**
- describe and explain the **conditions necessary for the successful formation** of coral reefs.

Coastal environments around the Caribbean vary in appearance and processes but have some common features. However, after the passage of Hurricane Ivan or some other storm event, the cliff line at a particular location you visited last year may have changed dramatically or even disappeared, or sand along a beach front may have been totally rearranged. The environment of most coasts is dynamic in nature, but others are more stable. The nature of coastlines is dependent on various factors, such as past tectonic processes, rock structure, slope of the land, and the energy of waves that strike the shore.

Wave types and characteristics

✳ **Waves** are the oscillating movement of the sea as wind blows over the water.

Winds are the dominant influence on the development of wave strength and wave direction. The stronger the wind, the more powerful waves will be. Storms far out at sea can influence the waves that break on distant shorelines. Strong ocean currents may also influence the nature of waves. As the wind moves over the open sea, swells or waves begin to form. The distance of open sea over which the wind (which creates a wave) blows before a wave strikes the coast is known as the **fetch**. Many waves are formed during the process.

The distance between two successive waves is the **wave length**, and the distance between the top or **crest** of a wave and the lowest point or trough is the **wave height** (Figure 7.1). As waves approach the shore they appear steeper and the oscillatory movement is more evident. At this point, where the sea bed rises towards the surface, the crest plunges forward on the shore and the wave form collapses.

The movement of water up the shore is known as the **swash**. Under the influence of gravity the water is immediately pulled back towards the sea – this is the **backwash**. Materials are moved along the coastline by a dynamic interaction between the swash and backwash, which remove, transport and deposit materials.

▼ *Figure 7.1 The different parts of waves*

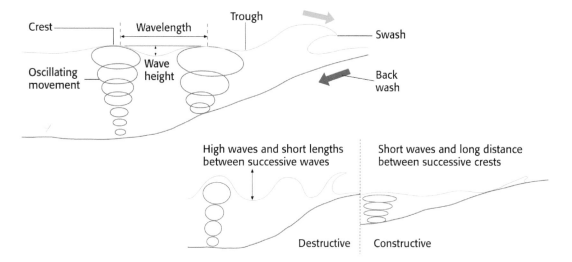

Wave forms influence wave movements and the way they break on shore. Waves that have short wave lengths and high wave heights, for example, tend to move faster and break with more force on the coast. The reverse occurs when the wave length is long and the height is low.

As waves break on the shore they affect the coastline. Waves are the principal forces behind the formation and modification of the coastline. Their impact on the coastline is dependent on various factors.

- **wave frequency:** rate per minute at which waves break on shore

- **wave height:** distance between the crest and trough

- **wave length:** distance between two consecutive crests

- **angle of wave crest:** direction at which waves approach the shore

- **steepness and gradient of land:** slope of the sea shelf as it approaches the shore.

Two types of wave are identified by these characteristics: **constructive** waves and **destructive** waves (Table 7.1).

➤ *Table 7.1 Features of constructive and destructive waves*

Constructive	Destructive
Long wave length 100m	Short wave length 20m
Low waves (little height) < 1m	Great vertical extent (high) > 1m
Less frequency per minute (< 12 per minute)	Greater frequency per minute (> 12)
Constructive to land (spread material)	Destructive to land (removes material)
Stronger swash, weaker backwash	Stronger backwash, weaker swash

▼ *Figure 7.2 Breaking wave on a Caribbean coastline*

Exercise

Refer to Figure 7.2 to answer the following.

1 Describe the wave forms in the photograph. Use the following terms: *oscillating, wave height, wave length, foam, swash, backwash.*

2 Describe the conditions that may have produced this wave.

3 Give the photograph a title or caption of your own.

Wave processes

Erosion

Wave action is an important agent of erosion. Materials are removed by various processes.

- **Corrasion** (also called **abrasion**): the scrubbing action of waves with sand and other fine material on rocks.

- **Hydraulic action:** the powerful effect of water in cracks and fissures, eroding rocks through compression.

- **Solution:** the chemical decomposition of limestone rocks by sea water.

- **Attrition:** the break-up of rocks that bounce and strike against each other as they move along the coast.

Transportation and deposition

Waves transport materials along the coast. These materials may have been eroded directly from the coastline nearby, but they may also have come from other locations, such as inland mountains, and been transported and deposited along the shore by rivers and the sea.

The swash of waves carries the materials and deposits them up the beach, while the backwash drags material back. By this process the materials are transported up and down the beach, and also moved sideways. This process of transportation is called **longshore drift** and is responsible for the development of most coastal features of deposition.

✳ **Longshore drift** is the movement of materials along the coastline by the action of breaking waves (swash and backwash), which generally approach the shoreline at an angle.

➤ *Figure 7.3 The process of longshore drift*

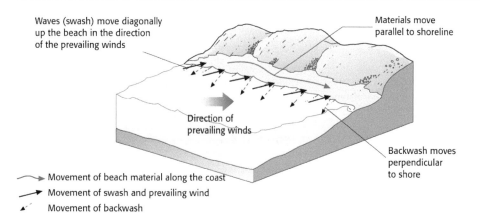

Waves (swash) move diagonally up the beach in the direction of the prevailing winds

Materials move parallel to shoreline

Direction of prevailing winds

Backwash moves perpendicular to shore

→ Movement of beach material along the coast
→ Movement of swash and prevailing wind
⤴ Movement of backwash

Coastal landforms

Coastal features of erosion

The effect of erosion is dependent on the structure and appearance of the rock. Soft rocks are less resistant than hard rocks and so are more susceptible to wave erosion. The most recognisable feature of coastal erosion is the presence of a cliff.

Cliffs are steep, often near-vertical slopes that rise abruptly from the sea. These are subjected to undercutting by hydraulic action and abrasion. As this erosion continues, a **notch** is cut into the base of the cliff. A cave develops, creating an overhanging cliff. As weathering and erosion continue the overlying section becomes weak and topples. Wave erosion removes the loose material which is used in further erosion. These features can be seen on many Caribbean coastlines such as the north coast of Trinidad and the east coast of Barbados.

▼ *Figure 7.4 Formation of cliff features*

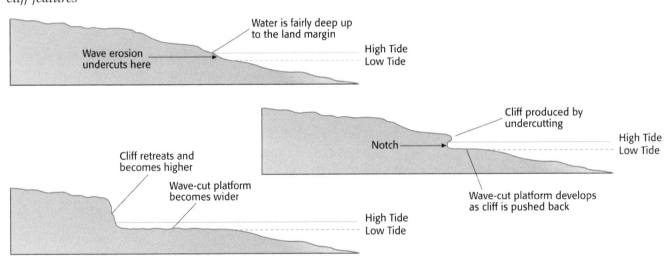

The notch is usually located between the high and low tide levels and is subjected to maximum erosion. Materials in suspension scour the base and the cliff recedes. As it does so, the effect of the wave on the cliff decreases. However, the abrasive action continues at the point where wave energy is highest. A platform replaces the retreating cliff. Wave action continues its scouring and cutting action on the platform, creating a flat-topped feature called a **wave-cut platform**. The platform is exposed at low tide and covered at high tide. It is sometimes loosely covered in beach material.

➤ *Figure 7.5 Features of a high-energy coastal environment*

Not all places along the coastline are equally vulnerable to erosion. Points of land that jut out into the sea come under attack from waves more than indented areas. These points are called **headlands** and the indentation between two headlands is called a **bay**. A headland develops when alternating layers of hard and soft rock are eroded at different rates. As waves approach a bay, they lose energy because of the reduced depth of the sea. The energy is therefore diverted towards the headland where cliffs are surrounded by deeper water. This process of deflection of wave energy around and towards the headland is called **wave refraction**.

➤ *Figure 7.6 Wave refraction around a headland*

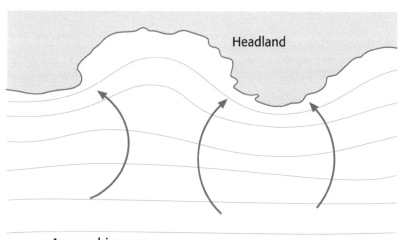

Approaching waves

As wave energy focuses on the headland, lines of weakness are attacked and the cliff face becomes disjointed. Fissures open under the influence of hydraulic action and corrosion, forming inlets and caves. If erosion occurs horizontally on both sides of a headland, the back walls of these caves may open up to form an **arch**. If the rate of erosion is accelerated by weathering the arch will collapse, leaving behind a **stack**. A stack is a pillar of rock (some may be as large as a small island) close to the shore which lies in direct line with the cliff and is made of the same materials.

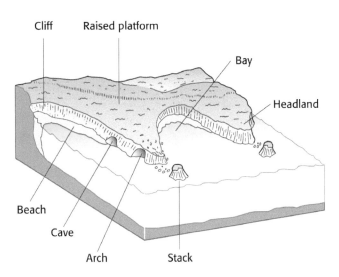

▲ *Figure 7.7 The evolution of cave, arch and stack features*

▲ *Figure 7.8 The coastal environment at Hectors River, Jamaica*

Exercise

1 Refer to Figure 7.8 to answer these questions.

  a Name the features being observed by the students in the photograph.

  b Explain the formation of each of these features.

  c What do you think could happen to the settlement above the cliff?

2 For a location in your territory, draw a sketch map of that section of the coastline.

  a Include features of erosion (cliff, wave-cut platform, headland and bay, cave, stack).

  b Draw a cross-section diagram for each feature.

Coastal features of deposition

When materials are transported and deposited along coastlines, several characteristic features may develop.

● The **beach** is the most common depositional feature. It is formed from the continuous accumulation of silt, sand, shingle and pebbles. However, it is also possible for large boulders to be deposited along a beach. Materials that make up beaches are derived from erosion of the land by both rivers and the sea. The material is deposited along the coast and longshore drift distributes the material. Most beaches in the Caribbean are sandy, but if a beach is composed more of shingle and pebbles than sand, it may be classified as a shingle beach. Where the coastline slopes very gradually into the sea, wide and sandy beaches are more likely to develop. Caribbean beaches vary in colour and composition from the black of volcanic islands to the white of coral.

➤ *Figure 7.9 A wide, sandy beach on the shore of Tobago*

- During a storm or high tide, large quantities of sand may be deposited up the beach, creating a parallel rise to the beach. If it stays in place, it may eventually become a **berm** or ridge of sand.

- A **spit** is formed when a large accumulation of material forms a narrow strip of land that juts out into the sea but is still connected to the mainland. Where a river carries large amounts of materials into a bay, waves moving obliquely will transport the material in a diagonal direction along the beach by the process of longshore drift. The spit may continue to grow, and if it joins onto an island or the mainland it becomes a feature known as a **tombolo**. Scott's Head in Dominica and Palisadoes in Jamaica are two well-known tombolos in the Caribbean.

- A **bar** is a narrow strip of sand and shingle formed by longshore drift but not connected to the mainland. Most bars are only exposed during low tide. It may also be called an **offshore bar**.

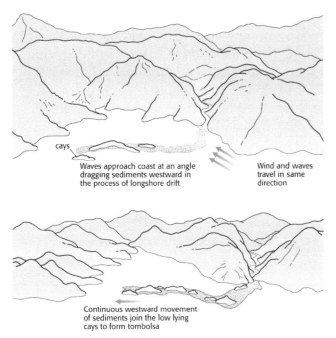

Waves approach coast at an angle dragging sediments westward in the process of longshore drift

Wind and waves travel in same direction

Continuous westward movement of sediments join the low lying cays to form tombolsa

cays

▲ *Figure 7.10 The development of a tombolo*

▲ *Figure 7.11 The Palisadoes tombolo in Kingston*

- A **mudflat** is made up mostly of silt and mud and usually develops behind a spit or near estuaries and swamps. Mudflats in the Caribbean are associated with mangroves and are important in the ecology of the marine environment.

➤ *Figure 7.12 Mudflats support the mangrove environment in Union Island, Grenadines*

Coral reefs

One very special coastal environment is the coral reef. These reefs form only under special conditions which create a distinct marine environment. Wave processes are modified along coral coastlines. There is a direct relationship between coral reefs and shoreline landforms.

Most people in the Caribbean know what the surface of a coral reef looks like, even if they are not really aware of its presence. There are coral reefs along the coastline of most Caribbean territories. When you look at the sea and observe waves breaking before they reach land, it is likely that there is a coral reef below the surface there.

✳ A **coral reef** is a community of marine animals in an ecologically balanced environment. The remains of the calcareous materials (shells) produced by corals and other marine organisms build up over time to form coral limestone.

Coral reefs are made from the accumulation of millions of tiny organisms called **coral polyps**. These tiny animals, which are similar to sea anemones, secrete calcium carbonate, creating a protective shell made of limestone in which they live. They can be delicately branched in various shapes and colours. Large numbers of them live together in colonies. When the polyps die, the shells are left behind for other polyps to build on. The shells from these and other marine animals build up over millions of years to form coral reefs. Today, the live coral on most reefs forms only a thin layer on top of a thick layer of older coral rock.

➤ *Figure 7.13 A coral polyp*

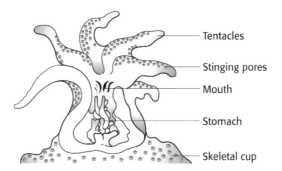

Tentacles

Stinging pores

Mouth

Stomach

Skeletal cup

A coral reef forms one of the most complex and intriguing ecosystems on Earth. It is often compared with the tropical rainforest environment. A wide diversity and great number of species (both plants and animals) live in this environment, displaying an array of brilliant colours.

➤ *Figure 7.14 Coral reefs are among some of the most diverse communities on Earth: they are called the 'rainforests of the ocean'*

Types of coral reef

Reef-building corals need a base on which to attach themselves. They cannot live in the deep ocean – they need something to build on. The polyps attach themselves to the undersea slopes of landmasses or islands.

There are two types of coral reef in the Caribbean, each found on a different type of landform.

- **Fringing reef:** this is the most common type in the Caribbean. The reef lies very close to the shore and the water between the reef and the shore is shallow. On an OS map it is marked by the symbol in Figure 7.16, which makes the reef appear to be attached to the land. Jamaica has an almost continuous fringing reef along its western, northern and eastern coastlines.

- **Barrier reef:** this lies further away from the shoreline and is separated from land by a deep lagoon. The Australian barrier reef is the largest barrier reef in the world (1 931 km). The barrier reef located off the coast of Belize is the second largest in the world, at 200 km long. (Smaller barrier reefs are found off the coasts of Antigua and the south coast of Jamaica.)

▼ *Figure 7.15 Symbol for a coral reef*

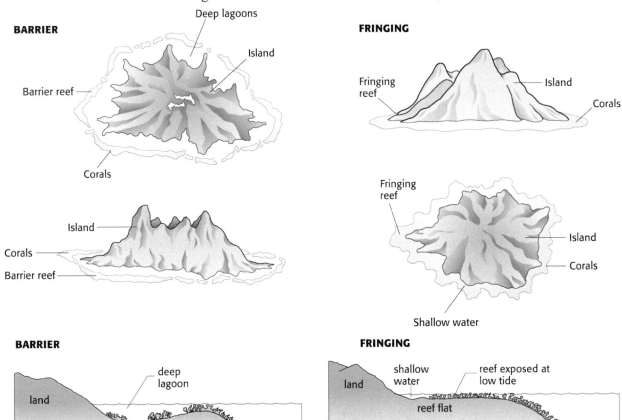

Figure 7.16 Types of coral reef

Many of the cays and offshore banks in the Caribbean are part of either a barrier or a fringing reef system that has risen above the present sea level. Many such reef banks are located off the coasts of the Bahamas, the Cayman Islands, Jamaica and Turks and Caicos. Most of these are covered at high tide or during storm surges.

➤ *Figure 7.17 The barrier reef off Belize*

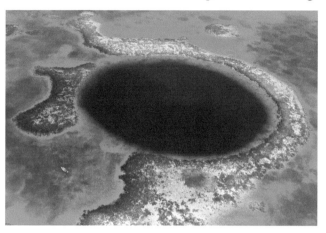

Location of coral reefs in the Caribbean

The warm, shallow waters off the Caribbean islands and over the neighbouring continental shelf mean that there is an abundance of coral reefs here – thousands of them. Every island has at least some coral reefs. These are a great attraction for tourists and a source of fish for local fishermen. The Bahamas, for example, has over 2 000 coral reefs – more than 700 of these are now raised above sea level.

▲ *Figure 7.18 Location of coral reefs in the Caribbean*

1 Using Figure 7.18 and your atlas, copy and complete the table below. You will need 10 rows, including Trinidad and Tobago, which has been done for you.

Territory	Area where reef is located	Type of reef	Name of reef
Trinidad and Tobago	South-west Tobago	Fringing	Buccoo Reef

  a Fill in each row with the name of a Caribbean territory.

  b For each territory name an area of coral reef and state the type of reef (fringing or barrier).

  c Where possible, name the particular reef.

2 a For one of the territories listed in your table, explain the importance of the reef named.

  b Draw a sketch a map showing the location of this reef in relation to the shore.

▲ *Figure 7.19 The good and bad conditions for coral reef growth*

The conditions needed for coral reef formation

Coral reefs will grow only where conditions are suitable. They must be in an area with a tropical climate. However, they need much more than just warm sun. The following are the ideal conditions for coral growth.

- **Warm seawater between 25° and 27°C:** cold water slows the growth but if the water is too warm it bleaches the corals, eventually killing them.

- **Sunlight is necessary for photosynthesis:** it allows the microscopic plants that corals feed on to make their food. These microscopic plants live in the tissue of coral polyps, where they get nutrients and protection. In exchange the corals have a ready supply of food.

- **The water must be clean, clear and well-oxygenated:** clear water enables sunlight to penetrate easily. Water that is polluted or carries a lot of sediment is not suitable. The sediment makes the water cloudy, preventing sunlight from penetrating the sea water. Sediments also choke the corals.

- **Corals grow best at a depth of 20–40 m:** this allows for the penetration of sunlight (light penetration becomes difficult below 100 m). The depth of water also affects temperature, as temperature decreases with depth of water.

- **Normal seawater salinity, together with gentle wave movement:** corals will not growth where the water is not salty enough, e.g. at the mouth of a river, or where it is too salty, e.g. in some parts of the Red Sea.

CHAPTER 8

Weather systems

By the end of the chapter students should be able to:

- explain the **differences between weather and climate**
- describe the weather associated with the **five main Caribbean weather systems**
- locate **areas in the Caribbean where these systems are dominant**
- explain how **relief produces variation in the climate of the Caribbean**.

The atmosphere is a very important part of our environment. It is composed of gases, such as nitrogen (78 per cent) and oxygen (20 per cent) with variable amounts of water vapour, carbon dioxide and ozone. There may also be small solid particles and pollutants.

Human beings are completely dependent on the atmospheric conditions, although we tend to take them for granted especially in the Caribbean. The atmospheric conditions of a place at a particular time are called **weather**. The atmospheric conditions we measure and record are: temperature, rainfall, wind speed and direction, and cloud cover and type. Atmospheric conditions can influence our health (e.g. asthma), transportation (land, air and sea), lives and property (e.g. hurricanes), and even the way we dress from day to day. Polluted air can result in climatic changes such as global warming (see Chapter 15).

> ✳ **Weather** is the actual conditions in the atmosphere at a particular place and time, usually over a 24-hour period.

> ✳ **Climate** is the average atmospheric conditions a place can expect based on records over 30 years. The climate of a place can be described as its average weather conditions.

Exercise

Indicate which of the following statements refer to climate and which ones refer to weather. Write down the number of the statement and put 'C' for climate or 'W' for weather next to it.

1   This morning is so hot but last night it was cool.

2   Guyana generally receives more rainfall than Antigua.

3   Rain in Guyana has stopped play in the second test match today.

4   The rainy season in Barbados usually ends in November.

5   It is too cool tonight to wear a strapless dress.

6   Take your umbrella, it is going to rain today.

7   There were many flight delays caused by the snowstorm in New York.

8   Sugar cane harvesting is done during the Caribbean's dry season.

9   The strong winds of Hurricane Ivan devastated Grenada.

10 Tourists get very sunburnt in the Caribbean sun.

The tourist season – December to April – is based on the difference in the climate, at that time, of the Caribbean, where it is hot, and of North America and Europe, where it is cold.

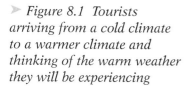
➤ *Figure 8.1 Tourists arriving from a cold climate to a warmer climate and thinking of the warm weather they will be experiencing*

Caribbean weather systems

The Caribbean countries are located between the equator and 30° north of the equator. They experience equatorial and tropical marine climates with hot, moist conditions. The main variation is in the seasonality of rainfall which is associated with the passage of weather systems. The Caribbean's weather is affected by five main weather systems: the Inter-Tropical Convergence Zone (ITCZ), tropical waves, hurricanes, cold fronts and anticyclones.

✳ A **weather system** is a large area of the atmosphere having special temperature and moisture conditions. Systems develop over one area and may move to affect other areas.

➤ *Figure 8.2 Distribution of weather systems in the Caribbean*

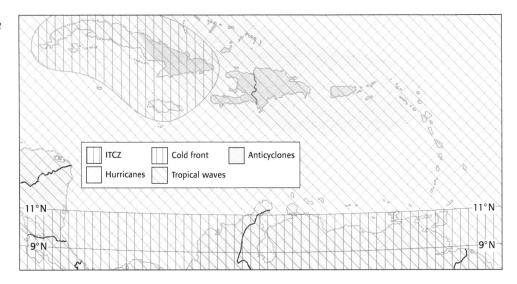

ITCZ	Cold front	Anticyclones
Hurricanes	Tropical waves	

The atmosphere in the Caribbean is hot and moist. Most Caribbean weather systems produce large amounts of rainfall as the hot air is forced to rise by convectional heating. The rising air exerts less pressure on the Earth's surface and may form a low-pressure system. Winds blow *into* low-pressure systems.

➤ *Figure 8.3 Convectional heating producing convectional rainfall*

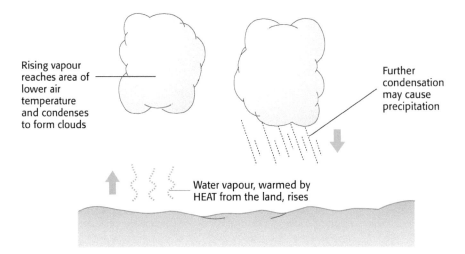

Rising vapour reaches area of lower air temperature and condenses to form clouds

Further condensation may cause precipitation

Water vapour, warmed by HEAT from the land, rises

During the dry season, the region may come under the influence of an anticyclone. This is a high pressure system with winds blowing out of it.

※ **Atmospheric pressure** is the 'weight' of a column of air. The atmosphere is held in place by gravity, and so it exerts pressure on the surface of the Earth.

Inter-Tropical Convergence Zone (ITCZ)

The ITCZ is an elongated area of low pressure lying east to west across the equatorial region. Hot air from the north (north-east trade winds) meets hot air from the south (south-east trade winds) in this area. The hot, moist air is forced to rise, resulting in heavy rainfall. As the air rises it creates an area of low pressure at the Earth's surface.

> *Figure 8.4 The ITCZ over Georgetown, Guyana*

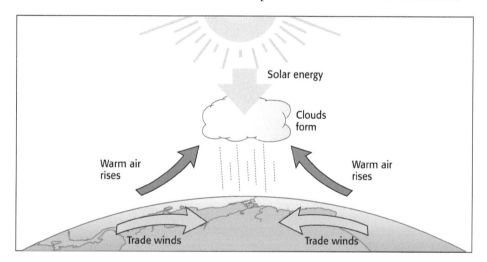

This belt of strong convection moves north and south of the equator with the seasons. In June, when the hottest place is north of the equator, the ITCZ drifts northwards and may affect Trinidad and Tobago and the islands of the southern Caribbean. In December, when the hottest place is south of the equator, the ITCZ drifts southwards and brings heavy rainfall to Guyana. It is very active in Guyana in March and September when the hottest place is at the equator. It may rain for many days at a time.

The ITCZ is associated with:

• heavy and continuous rainfall

• completely overcast skies with cumulus clouds

• winds that are light or calm

• temperatures remaining high.

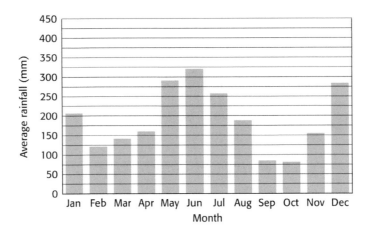

Figure 8.5 The relationship between the ITCZ and rainfall in Guyana

Exercise

Use Figure 8.5 to answer these questions.

1 What is the range of temperature?

2 Which two months are the wettest?

3 Are there any months without any rainfall?

4 What is the total rainfall for the year?

Tropical waves

These are weak low-pressure systems that do not have a closed centre. They are identified by the north–south axis of their pressure and show a northward bend in the isobars. They form near Africa over the warm waters of the Atlantic Ocean and move westwards into the Caribbean.

They affect most of the region except Guyana and Trinidad and Tobago during June to November. Many Caribbean countries such as Barbados are dependent on the passage of tropical waves to bring rainfall.

The passage of tropical waves is associated with:

● moderate winds

● heavy and continuous rainfall

● overcast skies

● temperatures remaining high.

Tropical waves differ from hurricanes in lacking a low pressure centre so they do not have very strong winds. However, tropical waves may develop into hurricanes.

➤ *Figure 8.6 Areas of rainfall in a tropical wave*

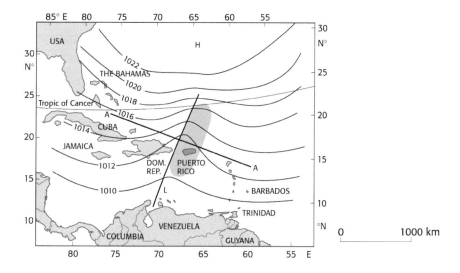

➤ *Figure 8.7 Weather associated with a tropical wave*

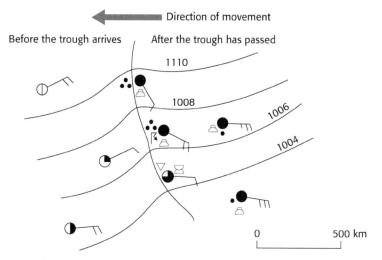

Hurricanes

Hurricanes are very strong low-pressure systems with strong winds (of more than 120 km/h – 75 mph) blowing into them in an anti-clockwise circulation, with heavy rainfall. They start as tropical waves near Africa and move westwards into the Caribbean during June to November. There is no single explanation for why some waves develop into storms and hurricanes and others do not. Many computer models have been used to try to predict the development of hurricanes.

Conditions favouring the development of hurricanes:

- warm water of tropical oceans (surface temperature of 26°C)

- Coriolis effect north and south of latitudes 10° North and South – this causes the spiral motion of the hurricane and influences its path

- divergent upper-level winds allowing updraughts of air.

Most Caribbean countries, including Belize and the Bahamas but excluding Guyana and Trinidad, are affected by hurricanes. Hurricanes are named each year, alternating between male and female names.

The main weather conditions associated with hurricanes in the Caribbean are:

- winds speeds above 120 km/h

- wind direction varying between north east and south west

- torrential rainfall

- skies completely overcast with towering cumulonimbus clouds

- thunderstorms with lightning

- temperatures remaining high, averaging 27°C.

The life of Hurricane Ivan

Hurricane Ivan started as a tropical wave in the Atlantic Ocean in early September 2004. Its curved its way through the Caribbean, strengthening over the warm waters of the Caribbean Sea. The sequence of events is given in Table 8.1.

➤ *Table 8.1 The passage of Hurricane Ivan*

Early September	Hurricane Ivan starts as a tropical wave in the Atlantic Ocean.
	Barbados is the first to be put on 'watch', and then on 'warning', as the most easterly of the Caribbean islands.
7 September	Hurricane Ivan passes south of Barbados.
9 September	It hits Grenada directly and drenches Tobago, then swings west-north-west.
12 September	Jamaica and the Cayman Islands are affected by gusts of 320 km/h, 8 m waves shifting the narrow tombolo of the Palisadoes landwards (see Chapter 7).
13 September	Western Cuba's tobacco fields are drenched.
17 September	Ivan goes ashore west-south-west of Montgomery in Alabama, USA, launching tornadoes, winds of 210 km/h and a 5 m storm surge.
24 September	Ivan is regenerated as a storm in the Gulf of Mexico.

Exercise

Use the information in Table 8.1 to plot Hurricane Ivan's path on a map of the Caribbean. Label your map with the dates.

➤ *Figure 8.8 Weather conditions during a hurricane*

These conditions are very hazardous to human life and property (see Chapter 14).

The development and movement of hurricanes is closely monitored by many agencies. You can go to the National Hurricane Center at www.nhc.noaa.gov and the National Weather Service at www.nwa.noaa.gov. These sites include official weather bulletins for the Atlantic Ocean area and the Caribbean. You can also find current weather maps and conditions for many places.

➤ *Figure 8.9 Sequence of weather during the passage of a hurricane*

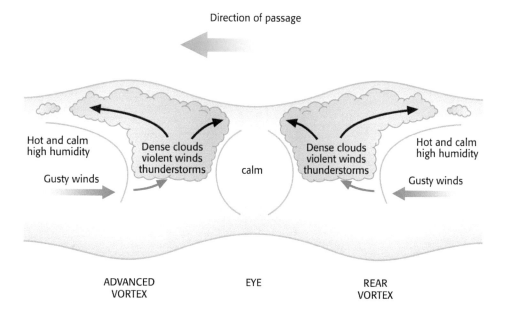

Exercise

Use Figure 8.9 to describe the sequence of weather on an island where the hurricane passes from east to west. Create a table with four headings: Before the hurricane, Advance vortex of hurricane, Eye, Rear vortex of hurricane.

Cold fronts

A front is an area where two air masses of different temperatures meet. The Caribbean weather is affected by cold fronts (sometimes called northers) which form over the northern Caribbean during winter. Cold, dry air moves south from the North American continent and pushes under the warm, moist Caribbean air. The rising warm air cools and condenses along a steep boundary, resulting in cumulonimbus cloud and heavy rain. This is called **frontal rainfall**. Behind the front, temperatures fall.

The Bahamas and the Greater Antilles, such as Jamaica, Cuba and Hispaniola, are affected by these weather systems (see Figure 8.2).

Weather associated with cold fronts:

● cooler temperatures, 18–21°C

● moderate rainfall that usually lasts for days depending on the movement of the frontal system

● moderate cloud cover (six-eighths of sky covered by stratus-type clouds)

● winds from the north.

➤ *Figure 8.10 Weather associated with a cold front*

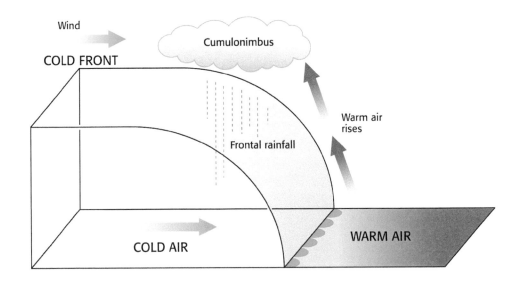

Anticyclones

Anticyclones are high-pressure systems. Winds blow out of them in an anti-clockwise direction. They form around the mid-latitudes (30°N) and may drift into the Caribbean. These areas are very stable with sinking air blowing out of the centre. Anticyclonic conditions can affect most of the Caribbean north of Trinidad and Tobago. (In some places these persistent winds are used for kite-flying competitions, around March–April).

Weather associated with anticyclones:

- dry, sunny
- clear skies
- moderate winds
- moderate temperatures.

➢ *Figure 8.11 Anticyclonic weather conditions*

System					
Weather characteristic	**ITCZ**	**Tropical wave**	**Hurricane**	**Cold front**	**Anticyclone**
Pressure	Low	Low	Very low	Low	High
Wind	Light	Moderate	Very fast (>75 mph)	Moderate	Moderate
Rainfall	Heavy	Heavy	Torrential	Moderate	Little
Cloud	Cumulus	Cumulus	Cumulus	Cumulus	No cloud
Temperature	28°C	27°C	27°C	21°C	30°C

⬦ *Table 8.2 Weather systems*

The influence of relief on climate

Relief is the height and slope of the land. The climates of the Caribbean are greatly modified by the presence of hilly or mountainous areas. Many countries of the Caribbean have fold or volcanic mountains, which rise high above sea level. These high areas affect temperatures and rainfall distribution of both the elevated area and surrounding areas.

Relief and temperatures

Generally, temperatures decrease as you climb higher above sea level. The air is cooler as heat from the Earth's surface is lost into space from the thinner atmosphere at high altitudes. An average 6.4°C is lost for every 1 000 m of elevation. Therefore the higher areas of the Caribbean, such as Blue Mountain peak in Jamaica at 2 229 m, will be 14°C cooler than Kingston, which is at sea level. Similarly in Dominica, Morn Diablotin at 1 419 m will be 9°C cooler than Portsmouth, which is at sea level. The low-lying Caribbean islands such as the Bahamas (highest point less than 200 m), Antigua (402 m) and Barbados (333 m) are not much affected greatly by relief – they are generally hotter and drier.

➤ *Figure 8.12 Differences in temperature of places at different altitudes*

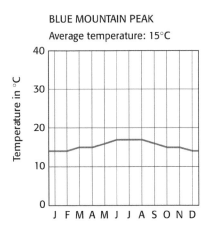

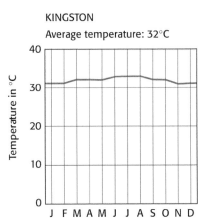

Relief and rainfall

Mountains actually create rainfall by forcing the air to rise over them. This type of rainfall is called **relief** or **orographic rainfall**. As the air rises it cools and condenses to form clouds. If the mountain is high enough it will rain on the side facing the wind or windward side. After it has lost its moisture, the dry air descends on the opposite or leeward side. This area does not receive much rainfall and is called the **rainshadow area**. This affects the distribution of rainfall in many Caribbean islands.

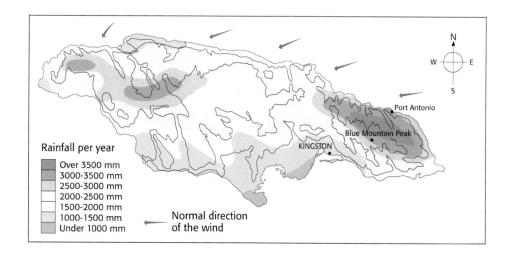

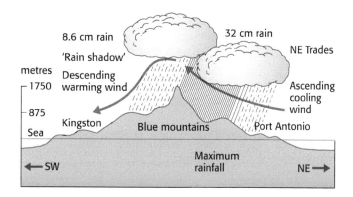

▲ *Figure 8.13 Relief rainfall in the Blue Mountains, Jamaica*

Exercise

Use Figure 8.13 to answer these questions.

1 What area receives the highest rainfall?

2 How much rainfall occurs in Kingston?

3 Compare the rainfall amounts for Port Antonio and Kingston. Give one reason for the difference.

CHAPTER 9

Climate, vegetation and soil

By the end of the chapter students should be able to:

- identify the **components of an ecosystem**
- describe the **characteristics of climate, vegetation and soil of the equatorial, tropical marine and tropical continental regions**
- locate **areas where tropical rainforests and tropical grasslands can be found**
- explain the **relationship between climate, vegetation and soils of equatorial and tropical continental regions**.

✳ An **ecosystem** is a biological (living) community within its own non-living environment. The two aspects are interdependent.

The word 'ecosystem' is really a contraction of the term 'ecological system'. **Ecology** is the study of the relationship between organisms and their environment, and a **system** is a set of interacting, organised parts. Ecosystems can be very small, for example a single tree may have a complex arrangement of interacting insects and plants in a particular soil and micro-climate. They can also be very large – the Amazon rainforest is one of the biggest ecosystems in the world, with many interdependent relationships.

▲ *Figure 9.1 Large and small ecosystems*

126

Human beings are considered such an important part of many ecosystems that their relationships with their environment are often studied as a separate subject – see Unit IV Human–Environment systems. People can have a huge impact on the environment (see Chapter 15) and even create their own environments (see Chapter 11).

Components of an ecosystem

An ecosystem can be divided into living and non-living parts.

- The **living** or **biotic** part consists of plants and animals.

- The **non-living** or **abiotic** part consists of soil and climate.

All components of an ecosystem have interdependent relationships with each other. The biotic elements include gaseous exchanges with the atmosphere, and chemical and nutrient exchanges with the soil.

Plants and animals

Most plant and animals depend on the energy from sunlight, although some special deep-ocean ecosystems get their energy from heat escaping from the Earth's interior. The plants and animals have different functions in the ecosystem.

- **Producers:** these are mainly plants that use energy to manufacture their food through photosynthesis.

- **Consumers:** these organisms get their food directly or indirectly from producers.

- **Decomposers:** these are special consumers which feed on organic waste.

The interactions of these living components result in the **net primary production** of an ecosystem (NPP).

➤ *Figure 9.2 Net primary production in an ecosystem*

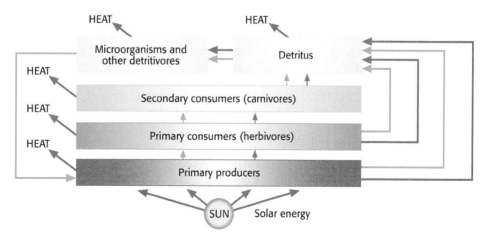

Soil

The Earth's surface is covered by a layer of weathered rocks (inorganic) and humus (organic). The humus is vital in processing inorganic material into nutrients which the plants can absorb. It is also important as a source of water for plants and to physically support their roots. The soil also interacts with the climate and biotic elements in the ecosystem. Rainwater infiltrates the soil and water vapour evaporates from its surface. Some plants fix minerals in the soil while burrowing animals help to aerate the soil. Dead vegetative matter creates the raw material of humus.

➤ *Figure 9.3 Soil*

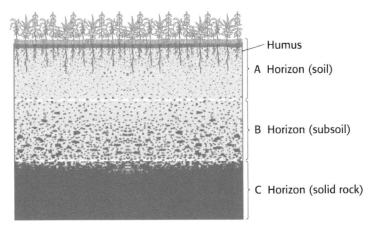

Atmosphere/climate

The atmosphere is the layer of gases around the Earth's surface on which living organisms depend. Most organisms use **energy from sunlight** to convert inorganic matter into organic matter (producers). Rain is also a very important source of **water** for life, since most living tissue is composed of a high percentage of water. Plants consume carbon dioxide and release **oxygen**, while animals may consume oxygen and release carbon dioxide. Plants transpire and release water vapour into the air.

➤ *Figure 9.4 Interdependent components of an ecosystem*

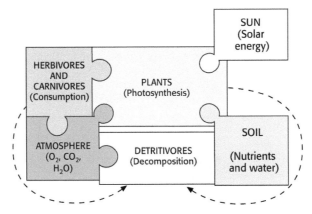

Climate, vegetation and soil of the equatorial, tropical marine and tropical continental regions

There are two important elements of climate: temperature and rainfall. Variations in these elements result in different climatic types.

- **By temperature:** we can identify **equatorial** and **tropical** regions as having temperatures which are hot or warm all year (temperate and polar regions are cold for part of the year).

- **By rainfall:** we can distinguish **marine** climates (near the sea) from **continental** climates (far from the sea). Marine climates are generally wetter than continental ones. The marine areas receive moist winds off the sea which increase their rainfall. In the continental areas, the winds are dry, so these areas have lower rainfall.

In the Caribbean we experience the **equatorial** climate, **tropical marine** climate and **tropical continental** climate. These climates are affected by the passage of many weather systems (see Chapter 8).

➤ *Table 9.1 Climates of the Caribbean*

Climate	Countries	Weather systems
Equatorial	Guyana, Trinidad and Tobago	Inter-Tropical Convergence Zone
Tropical marine	Islands in the Caribbean Sea	Hurricanes, tropical waves, anticyclones
Tropical continental	Southern Guyana	Continental high pressure system

The climate of an ecosystem affects both its vegetation and its soils. Ecosystems are therefore often identified by their climate type.

➤ *Figure 9.5 Distribution of world climate and vegetation*

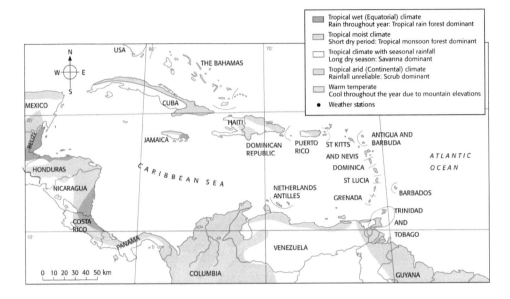

Climate, vegetation and soil of the equatorial regions

Climate

The distinct climatic features of equatorial regions are:

- total annual rainfall more than 2 000 mm – daily afternoon showers

- constant high temperatures – 26–29°C, annual range 3°C

- heavy rains of the ITCZ in March–April and September–October

- high humidity.

Exercise

The average temperatures and rainfall for Georgetown, Guyana, in the Caribbean equatorial region are as follows.

	J	F	M	A	M	J	J	A	S	O	N	D
Temperature (°C)	26	27	28	28	27	26	26	25	25	25	25	26
Rainfall (mm)	5	10	18	28	13	13	13	10	8	8	5	3

1 Draw a line graph to show the temperatures, and a bar chart of the rainfall, for each month for Georgetown. (See Chapter 2 for detailed instructions on how to do this.)

2 Describe the climate shown on the graph by stating each of the following:

- annual maximum temperature
- annual minimum temperature
- annual range of temperature (maximum minus minimum)
- total rainfall (add together the amounts for each month)
- seasonality of rainfall – rain all year/summer maximum or winter maximum/wettest month/s.

The equatorial climate is found around the equator, and in the Amazon basin (Brazil, Guyana), the Congo basin (Democratic Republic of the Congo) and the islands of South-East Asia (Malaysia, Indonesia) – see Figure 9.5.

Vegetation and soil

The vegetation and soil of an ecosystem are closely interrelated. The vegetation is tropical rainforest and the soil is latosol. The trees of the rainforest protect the soil from the hot, wet climate and contribute dead leaves to the soil organic matter (humus). The soil provides nutrients, water and support for the trees.

The **vegetation** of the equatorial region has some characteristic features.

- The dense layer of tall trees forms a canopy over the top, with less dense layers of trees below it. These are ideal conditions for plant growth (hot and wet), although the trees have to compete for light.

- The rainforest has an evergreen appearance, with a year-round growing season; individual trees tend to drop their leaves throughout the year.

- The trees have:
  - large leaves with drip tips to let the water run off easily
  - flattened crowns branching near the top
  - very tall trunks up to 20 m in height
  - thin bark
  - large buttress roots to support their tall trunks
  - shallow roots spreading through the topsoil.

- Many species of plants and animals live together.

- There is little undergrowth because of the lack of sunlight on the forest floor.

- Many vines and parasitic plants live on the trees to reach sunlight and/or for nutrients.

- There is a lot of leaf litter on the forest floor as the trees grow continuously in this seasonless climate.

- Many fungi and mosses live on rotting organic matter on the ground.

The soils are **latosols**, which lie below the rainforest and are dependent on the trees for shelter from heavy rainfall to maintain their structure. These soils are very prone to soil erosion if they are exposed to the elements. The soil:

- extends to great depths, with weathering occurring as deep as 20 m

- has a thin humus-rich topsoil as decay is rapid in these hot wet conditions. This is maintained by the accumulation of leaf litter from continuously shedding trees

- is strongly leached by the water from the daily showers of rain

- has iron and aluminum oxide deposits created by the leaching of silica

- is home to many living organisms which break down the complex inorganic compounds so the trees can absorb them.

✳ **Leaching** is the downward movement of soil water transporting soluble minerals into lower layers. In the equatorial forest, the soil water may drain into the groundwater and flow out of the soil.

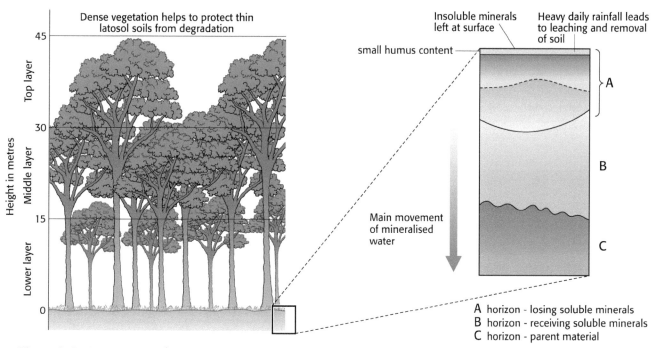

▲ *Figure 9.6 Vegetation and soil features of the equatorial rainforest*

➤ *Figure 9.7 Equatorial and Tropical Continental regions of Guyana*

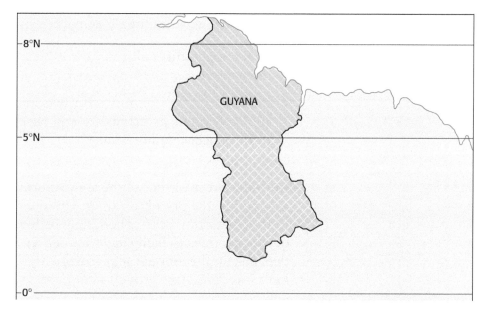

Guyana is the largest Caribbean territory, with an area of 215 000 km². It occupies part of South America and lies between 2° and 8° north of the equator. Large parts of the country experience an equatorial climate which supports dense rainforest.

Go to www.forestry.gov.gy/forests.htm for more information on forests in Guyana.

Exercise

Use Figure 9.7, or a map in an atlas, to answer the following question.

1 Describe the distribution of forest. Explain why it is located in this part of Guyana.

Climate, vegetation and soil of tropical marine regions

Climate

The climate experienced by tropical marine areas is very varied. Coastal areas of islands in tropical latitudes receive varying amounts of rainfall. All the islands in the Caribbean Sea have a tropical marine climate. Trinidad and Tobago, at latitude 10°N, have a wetter marine climate; while Jamaica and Cuba, near 23°N, experience a distinct dry season.

The tropical marine climate is marked by:

● high temperatures all year, 28–30°C, with a small annual range of temperature

● rainfall 1 000–1 500 mm, with seasonal variation (generally with a July to September maximum and experiencing tropical disturbances and hurricanes).

Vegetation and soil

In wetter areas of the southern Caribbean and on the higher mountains, the broadleaf evergreen forest **vegetation** is similar to that in the equatorial areas. There may be fewer tree layers and more deciduous trees (which shed their leaves in the dry season). This depends on the amount and seasonality of the rainfall. Drier areas have less dense tree growth with more grasses and shrubs.

The **soil** varies widely, depending on the parent material and rainfall regime. Some soils in the tropical marine climate are latosols with strong leaching, while others experience **calcification** during the dry season.

Climate, vegetation and soil of tropical continental regions

Climate

These areas, in the interior of continents, are characterised by:

- high temperatures all year, varying between very hot (over 30°C) in the dry season to hot (21°C), resulting in a larger annual range of temperatures than in the equatorial and marine climates – there are no moderating sea breezes in these areas

- a moderate total annual rainfall (500–1 000 mm)

- a distinct dry season lasting 4–7 months.

The tropical continental climate is experienced in the centre of continents between 5° and 15° north and south of the equator – that is, north of the equatorial regions and inland of the marine regions. These inland continental areas include parts of Brazil, Venezuela and Guyana, Central and East Africa and areas of northern Australia.

Few areas in the Caribbean experience this climate, but one such area is southern Guyana. The Rupununi savannah is in south-west Guyana. It covers an area of 15 542 km². It is mainly open woodland, consisting of tall grasses with some trees, and experiences distinct wet and dry seasons.

Vegetation and soil

The **vegetation** of the tropical continental regions is tropical grassland. This has several characteristic features.

- Coarse grasses, such as elephant grass, thrive in this climate. They have:
  - long sheaves
  - deep roots
  - a deciduous habit (they wither and become dormant in the dry season)
  - a height of 3–5 m in the rainy season.

- Specially adapted trees such as acacia and baobab may be scattered in wetter areas. The baobab has:
  - a small flattened crown with few, thin branches
  - a huge trunk to store water, with a diameter up to 10 m
  - medium height
  - long extensive roots to tap into groundwater
  - small thin leaves to reduce transpiration in the dry season.

Go to www.wwfguianas.org/ecoreg_savanna.htm for more about the savanna regions of Guyana.

The **soil** of tropical continental areas is generally not very deep, its depth depending on the amount of rainfall, location and parent material. Wind-borne minerals may be added from adjacent deserts. The deciduous grasses whither and accumulate as a thin dark layer of humus. Decay is slow during the dry season. The texture is good and loose except where continuous wetting and drying has hardened the exposed laterite layer. Leaching occurs during the wet season, leaving behind iron and aluminum oxides. In the dry season calcification may leave mineral nodules in the B horizon. These soils are ferruginous.

The vegetation and soil of the tropical continental areas are closely related. The grasses die and add humus to the soil, while the tufted grasses hold the soil together and their roots penetrate deep into the soil in search of moisture.

➤ *Figure 9.8 Interdependent tropical continental climate, grassland and soils*

Deciduous grasses, up to 6m high during wet season

Soil: leaching occurs in wet season; calcification may occur in dry season

Deep roots stabilise soil. Vegetation withers to form thin layer of humus.

Exercise

1 Describe the tropical continental climate.

2 Describe two ways in which the climate influences:

a the vegetation

b the soil of an area.

3 Describe one way in which the grassland vegetation influences the soil characteristics.

Exercise

Identify and describe four differences between the climate and vegetation in the wet and dry seasons in a tropical continental region.

Exam practice questions

Natural systems

Paper 1: Multiple choice

1 Which of the following features is a product of volcanic activity?
   A Anticline
   B Crater
   C Atoll
   D Delta

2 Which statement is true of a lava plateau?
   A It has ash and lava
   B It erupts violently
   C It produces a lot of ash
   D It produces gently sloping lava

3 Features associated with dying volcanic action are:
   A Hot springs
   B Tsunamis
   C Nuée ardentes
   D Lava flows

Use Figure P.1 below to answer Questions 4–6.

 Figure P.1

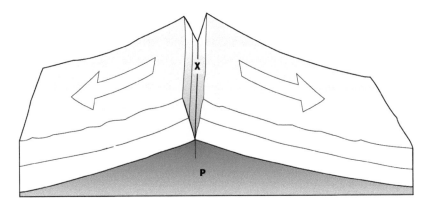

4 The area labelled P is the:
   A Core
   B Plate
   C Welling zone
   D Mantle

5   The pair of arrows show that this is a margin where plates are:
     A Converging
     B Moving neutrally
     C Diverging
     D Being cracked

6   The zone marked X is an area of:
     A Subduction
     B Upwelling
     C Faulting
     D Subsidence

7   Denudation is best described as:
     A Weathering and land forms
     B Erosion and weathering
     C Stripping and slumping
     D Carbonation and erosion

8   Which of the following is a process of mass wasting?
     A Soil creep
     B Frost action
     C Carbonation
     D Pot-holing

9   Temperature decreases with increasing altitude in the atmosphere, because the air is heated:
     A Directly from the sun
     B By heating of the clouds
     C From the earth's surface
     D From the upper layers

10   The climate characterised by an annual temperature range of about 10°C and rainfall usually below 1 000 mm annually is:
     A Tropical marine
     B Tropical continental
     C Tropical monsoon
     D Tropical rainy

11   The vegetation associated with a tropical continental climate is:
     A Scrub
     B Rainforest
     C Grasslands
     D Trees

12   The only barrier reef in the Caribbean is in:
     A Jamaica
     B Guyana
     C Belize
     D Tobago

13 A dendritic drainage pattern is one in which the rivers form a:
   A Grid square
   B Tree-like shape
   C Large channel
   D Circle

14 Karst limestone landscape has all the following except:
   A Swallow-holes
   B Underground caves
   C Mass movement
   D Dolines

15 A landslide is:
   A A fast-moving river
   B A fast mass-movement
   C A slow soil movement
   D Creeping land

Total 15 marks

Paper 2

2 (a) Draw a well-labelled diagram to show plate movement at a divergent plate boundary. Include arrows to show movement, and features produced by the movement.   (4 marks)

   (b) (i) What is 'intrusive volcanic activity'?   (2 marks)

        (ii) Name and describe two features of intrusive volcanic activity.   (6 marks)

   (c) Explain carefully two factors that influence the characteristics of volcanic eruptions.   (6 marks)

   (d) Explain two changes that may occur in volcanic landforms over time.   (6 marks)

Total 24 marks

3 (a) The diagram below shows a cross-section through a river channel. Use it to answer the questions below.

Figure P.2

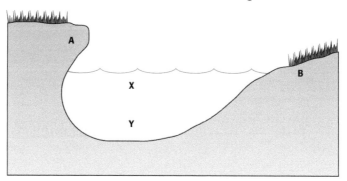

(i)   Name the landforms labelled A and B.

(ii) Is the water flowing fastest at point X or point Y?

(iii) In what part of the course is the channel likely to look like this diagram?   (4 marks)

(b) (i)   Name three different types of weathering.   (3 marks)

(ii) Describe two specific weathering processes and the result of each process.   (4 marks)

(c) (i)   Explain how materials are moved along the coast by wave action.   (3 marks)

(ii) Name one landform produced by the process explained in (i).   (1 mark)

(d) Explain how any three of the following landforms are produced.

• wave-cut platform   • ox-bow lakes   • underground caves
• deltas   • cockpits   • beaches   (9 marks)

Total 24 marks

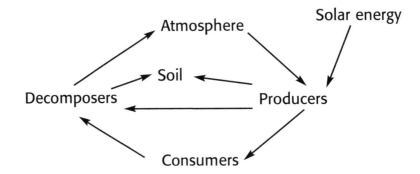

4 (a) Use the diagram above to answer the following questions.

   (i)   What process connects rocks and soil?

   (ii) Name two components of an ecosystem that are influenced by climate.

   (iii) How are plants and rocks connected?   (4 marks)

(b) (i) Describe two characteristics of the soil of tropical continental regions.   (4 marks)

   (ii) Describe two tree layers of the tropical rainforest.

                                                         (4 marks)

(c) Explain two ways in which the vegetation and soil of the tropical rainforest are interdependent.   (6 marks)

(d) Compare the tropical marine and tropical continental climates under these headings:

'Annual range of temperature' and 'Seasonality of rainfall'.

                                                         (6 marks)

Total 24 marks

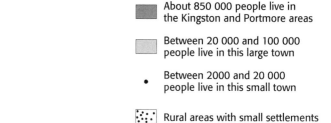

CHAPTER 10

Population

WHERE PEOPLE LIVE

By the end of the chapter students should be able to:

- explain the **factors influencing distribution and density of population** in one Caribbean country
- compare the **factors affecting the growth of population** in one Caribbean country and one developed country.

key

About 850 000 people live in the Kingston and Portmore areas

Between 20 000 and 100 000 people live in this large town

• Between 2000 and 20 000 people live in this small town

Rural areas with small settlements

▲ *Figure 10.1 Distribution of population in Jamaica*

The world's population stands at 6.4 billion, according to the 2004 *Population Reference Bureau Report*. Approximately 40 MILLION of that total live in the Caribbean.

✳ **Population** refers to the total number of people living in a given place at a given time.

✳ **Population distribution** describes the way people are **spread out** over a given region. The global distribution is uneven.

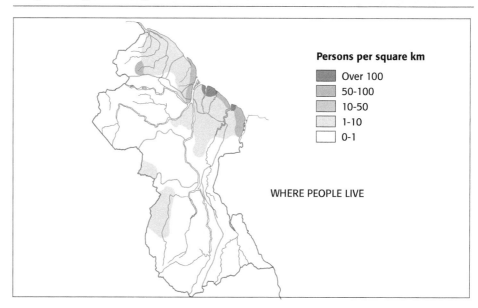

Persons per square km

Over 100

50-100

10-50

1-10

0-1

WHERE PEOPLE LIVE

▲ *Figure 10.2 Density of population in Guyana*

143

Figures 10.1 and 10.2 show how population is spread out over the territories of Jamaica and Guyana. Note the areas that have more people and the areas with fewer people. In each of these territories the population is spread consistently along the coast but is sparse to non-existent in parts of the interior. This is typical of many Caribbean countries.

✳ **Population density** refers to the number of people in a square unit of area in a given place.

Population density is usually calculated in square kilometres (km²) or square miles (sq. mls). To calculate density, take the population of a given place and divide it by the area of that place.

Examples

The population of Jamaica is approximately 2 772 900.

The size of Jamaica is 10 991 km².

The population density of Jamaica is approximately **252** people per square kilometre.

The population of Kingston (capital city of Jamaica) is approximately 96 052.

The size of the city is 22 km².

The population density of Kingston City is approximately **4 366** persons per square kilometre.

The population of Trelawny parish is approximately 74 060.

The size of the parish is 875 km².

The population density of Trelawny is approximately **85** people per square kilometre.

Exercise

Why is there such a great difference between the population densities of Kingston and Trelawny? Discuss.

Remember, population **distribution** refers to *spread*, while population **density** refers to *concentration* of people.

➤ *Figure 10.3 Features of dense population in West Kingston*

Factors influencing population distribution in the Caribbean

Historically, certain factors have influenced population distribution in the Caribbean. The region was first settled by Amerindians. They had a simple lifestyle and so they settled where there were basic resources such as water, and good soil for planting crops. Many settled by the sea because fishing was their main activity.

When the Europeans arrived with Columbus in the late 15th century, the Caribbean islands were undeveloped in every way – the Bahamas and Barbados had forests and wetlands. The European settlers needed more resources than the Amerindians – these needs were driven by their more advanced economic development. For these people the factors influencing distribution of population were different.

▼ *Table 10.1 Factors influencing population distribution and density in the Caribbean*

Factors	Positive (Dense)	Examples	Negative (Sparse)	Examples
Relief (physical)	Flat land, undulating land, coastal lowlands	• Barbados (South) • Belize (Belmopan, Corozal) • Bahamas	High and steep terrain	• Belize (Maya Mts) • Jamaica (Blue Mountains)
	Valleys	• St Vincent (Buccament, Cumberland) • Dominica • Trinidad (Port-of-Spain) • Jamaica (Richmond)	Deep ravines, rugged terrain	• St Vincent (Soufrière) • Jamaica (Blue Mountains, Cockpit Country)
	Foothills of volcanoes	• St Lucia (Soufrière) • Montserrat • St Kitts • Martinique	Hazardous	
Vegetation and soil (physical)	Deep, fertile soil	• Belize (Orange Walk) • Guyana (Rupununi Savanna) • Jamaica (southern parishes) • Trinidad (Toco, Savana Grande)	Thin soils in mountainous regions, highly leached and lacking humus, natural forests	• Barbados (Mt Hillaby) • Belize (Maya Mts) • Dominica (Mount Trios) • Guyana (Pakaraima Mts) • Jamaica (New Castle, Mecca Sucker – Blue Mountains) • St Lucia (Mount Gimie, Quilesse Reserve) • St Vincent (Soufrière) • Trinidad (Northern Range)
	Volcanic soil	• Dominica (Dublanco, Salisbury, St Patrick) • St Lucia (Dennery, Micoud, Laborie) • St Vincent (Colonarie, Stubbs, Calliaqua) • Montserrat (Plymouth)	Poorly drained, swampy	• Belize (Stann Creek) • Jamaica (Ferry, Great Morass – St Thomas)

Factors	Positive (Dense)	Examples	Negative (Sparse)	Examples
Natural resources (physical)	Minerals	• Guyana (Kurupukai – gold) • Jamaica (Manchester – bauxite)		
	Oil	• Trinidad (west)		
Water supply (physical)	Surface and underground supplies	• Barbados (Bridgetown) • Belize (Belmopan, Belize City) • Grenada (all round) • Guyana (Georgetown) • Jamaica (Kingston) • St Lucia (all round) • St Vincent (all round) • Trinidad (Port-of-Spain)	Unreliable supplies	• Jamaica (Cockpit Country, Portland Point)
Economic (human)	Deep harbour (ports)	• Bridgetown • Castries • Soufrière • Kingston • Georgetown • Port-of-Spain	Shallow harbour	• Belize City (like many Caribbean ports, the harbour had to be dredged)
	Tourism	• Jamaica (Ocho Rios, Montego Bay, Runaway Bay, Discovery Bay, Negril, Treasure Beach, Port Antonia) • Barbados (Worthing, Bridgetown, the west coast) • St Lucia (Castries, Gros Islet) • Bahamas (Nassau, Grand Bahama – Freeport City)		
Urban (human)	Public and social services, transport links, industrial and commercial facilities	All Caribbean capitals have dense population	Overcrowding (resulting in some people moving out)	• Belize City • Kingston • Port-of-Spain

Note: Positive factors encourage dense population while negative factors contribute to sparse settlement.

▲ *Figure 10.4a Town of Soufrière, St Lucia, in valley where fertile volcanic soil encourages farming*

▲ *Figure 10.4b Buccament Valley in St Vincent – surrounding hills shelter settlement from strong winds*

Factors that influence population distribution in Jamaica

- **Relief/topography:** flat and undulating lands are ideal for settlement. Most settlements in Jamaica, especially the towns, are on low land. With the exception of Mandeville, all parish capitals are on low-lying plains or along the narrow coastal strip around the island. The towns have the largest populations. Kingston, the largest settlement in Jamaica, is on the Liguanea Plain at the foot of the Blue Mountains. Like many other Caribbean territories, Jamaica is hilly, with some areas being very high (over 2 000 m). Settlement was difficult for early settlers, as the hills were treacherous and difficult to traverse. Even with today's heavy equipment, like excavators and power cranes, some areas are still difficult to cut roads through. The Blue Mountains and John Crow mountain region in the east of the island are not only high but also steep. These areas are very susceptible to slope failure (landslips). The Cockpit Country in the centre of the island is also precipitous. The weathered sinkholes, caves and escarpment mean that road construction and movement of equipment are difficult (see Chapter 5). Settlement in these areas will always be difficult and so they will not attract population. Look back to Figure 10.1 to see the distribution of population in these areas.

- **Water supply:** 'water is life' – this is a frequently used slogan of the National Water Commission of Jamaica, because water is indeed one of the basic essentials for survival, for both plants and animals. Before technology brought us pumping stations and piped

➤ *Figure 10.5 Kingston sprawls across the Liguanea Plain*

water, people settled near rivers and springs where they had easy access to water. Many settlements in Jamaica sprang up beside rivers and springs and took the name of the local water source, for example Constant Spring and Golden Spring in the parish of St Andrew. Others include Black River (a parish capital), Rock Spring, Ulster Spring, Mason River, Rock River, White River and Millbank along the Rio Grande River. These are some of the larger settlements, but many smaller settlements are also named after water sources. The capital city of Kingston is near the Hope River and today receives water not only from the Hope River but also from the Yallahs River and Wagwater River.

- **Soil:** originally, for building a settlement the local soil needed to be relatively porous, not heavy clay and not too sandy. However, technology has changed this. Jamaica had many mangrove swamp areas, but it is now possible to create new building land from these swamps. Portmore and the industrial areas along Kingston Harbour have been filled in with debris and topsoil and then built over. Such areas are not suitable for farming. In those areas where agriculture is dominant, the land is well drained. Flat land, good drainage and fertile soils are ideal for agriculture. The large sugar plantations occupied such land during the colonial period, in areas such as Bushy Park and Innswood on the St Dorothy Plain in St Catherine, and Hope and Mona on the Liguanea Plain. There are large settlements in all these locations today.

- **Natural resources (minerals):** some areas in Jamaica are rich in bauxite ore. The town of Mandeville in the hills of Manchester developed as a result of local bauxite mining. Today, though, bauxite production is significantly reduced, and many of the hotels and apartments in the town that once housed expatriates are now offices. In other areas where aggregates like limestone and gypsum

▲ *Figure 10.6*
Bauxite Plant

are mined, small communities sprang up around the activity – for example Bull Bay in St Thomas – but immediately outside the area of mining the population is generally sparse. Some of these communities have continued to grow despite the reduction in mining and excavation activities.

- **Access to the coast:** most Jamaican towns are situated along the coast, especially in the north. The earliest settlers – the Tainos (formerly called Arawaks) – were hunter-gatherers and fishermen, and they mostly settled along the coast. Discovery Bay and Runaway Bay are old Taino settlements that have grown into large modern settlements. The Europeans also settled along the coast and developed trading posts. Jamaica has many deepwater ports because of the natural harbours on its coast. Kingston is the seventh largest natural harbour in the world. The coastline also influences population distribution, with the development of tourism. The concept of clear blue waters and white sandy beaches is still a feature of tourism development, especially along the north coast. Montego Bay is the second largest population centre in Jamaica and is also the largest tourist resort on the island.

- **Political and social:** some communities sprang up as political strongholds. These communities started as squatter settlements and then over time were accepted by governments because of their political support and because amenities such as utilities, schools and health centres were established. Operation Pride, a government housing development programme between 2000 and 2003, established new communities as a part of a social reform programme to provide housing. The site of such places is usually dependent on availability of land and not necessarily on any particular physical factor.

The factors listed above are not the only ones influencing population distribution in Jamaica, but they are the factors that have contributed most to the distribution of older settlements. However, the pressures of population growth and technology have influenced the spread of population into other less conventional areas.

Exercise

1 Using an atlas, draw sketch maps of three Caribbean territories. Include the major towns on your maps.

2 Using Table 10.1 and an atlas, compare the population distribution of these countries with that of Jamaica.

Population growth and change

Population in the Caribbean is constantly changing. People are never static but always moving, and there is continuous growth.

The population of the world has seen an alarming increase since the start of the nineteenth century. Can you say what happened in the nineteenth century to cause this rapid increase?

➤ *Figure 10.7 World population growth over the last 1 000 years*

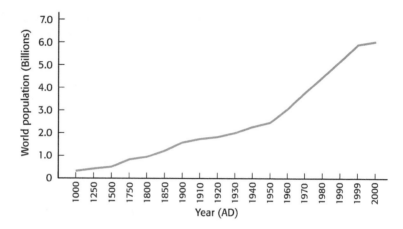

The population grew so fast that people began to fear that the world could not sustain so many people. Over the past 20–30 years, therefore, great efforts have been made by many countries to reduce the growth of their population. However, some countries still have a large, growing population, and across the globe there is a very uneven distribution.

Exercise

Referring to Figure 10.7, list the regions that have had a rapid population growth over the last 100 years.

Population growth is the increase in a given population over a period of time – this could be 1 year or 10 years. Many countries carry out a census (counting of people in a country) every 10 years.

Population change is the increase or decrease in the total number of people at a place over a period of time.

Population growth or change in a country depends on three factors:

- **births** (number of babies being born)

- **deaths** (number of people dying)

- **migration** (number of people leaving or entering a country permanently).

❋ **Calculating population change of a country**
Last count + Births – Deaths ± Migration = Population

Note: migration is not a factor when calculating world population.

Natural increase is an increase in population resulting from more births than deaths.

It is difficult to evaluate population using all these figures. So demographers refer to **birth rates** and **death rates**, which measure births and deaths per 1 000 people, usually over a period of one year.

Calculating birth and death rates

Birth rate   =   total number of births/total population x 1 000

Death rate   =   total number of deaths/total population x 1 000

These rates help us to understand the frequency and extent of change taking place. For example, if there were 1 000 deaths in Barbados, which has a population of approximately 272 000, and the same number of deaths in Trinidad, with an approximate population of 1 162 000, the implication for the two countries would be very different: the death rate would be very much higher in Barbados, because it has a smaller total population. (You can calculate the actual rate using the formula above.)

Changes in population do not just happen; they are influenced by social, political and economic conditions occurring in a place. Caribbean countries are at different stages of development: some are more wealthy than others, social conditions vary from place to place, and some have serious political instability. Haiti, for example, has had several periods of political turmoil and is the poorest Caribbean country. In chapter 11 (pages 167–168) you can evaluate the population growth rate for Haiti and come to your own conclusion. Collectively, however, Caribbean countries are working hard to reduce the rate of population growth.

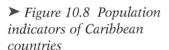

➤ *Figure 10.8 Population indicators of Caribbean countries*

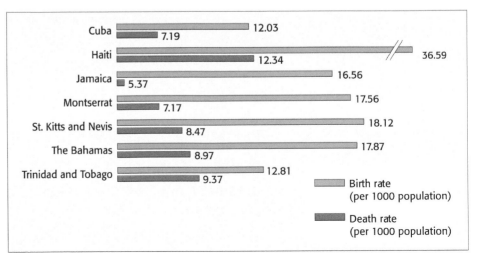

Factors affecting population growth in Jamaica

Social

Improved education: It is only through public education and a sound population policy that population growth can be checked. Past ideas about 'having out their lot' have contributed to high fertility rates. Jamaica has had successful population policies that have contributed to a decline in its birth rate. Barbados has the highest literacy rates in the Caribbean and one of the lowest birth rates. Education empowers women to care for their health, makes them employable and gives them choices.

Teenage pregnancy: The younger a woman is when she starts having children, the more children she is likely to have, because she has more child-bearing years. Much has been done to control the rate of teenage pregnancy through programmes like Youth Now. This was a reproductive health programme targeting teenagers to encourage them to practise safe and responsible sex through the use of condoms. If teenage pregnancy is controlled and young girls get an education and a career, it is thought that they are less likely to have many children.

Economic dependence of women on men: This dependency affects women primarily in the lower socio-economic sectors. Poverty forces them to depend on men for economic survival. The man demands that she has a child for him; when he leaves she turns to another man for support for her and the child; he demands of her to have his own child, after which he too may leave and the vicious cycle continues. This social problem is one that affects the birth rate in Jamaica.

Economic

Poverty: This is the driving force behind high birth and death rates in developing countries. Poverty prevents access to all the amenities for good living such as healthcare, education and shelter. In Haiti, where over 50% of the population live in poverty, death rates are high, especially among infants and old people. Children die from disease and malnutrition. But the population still grows as women tend to have more children under these conditions. In Jamaica, the government has embarked on poverty alleviation programmes through agencies like USAID and UNESCO.

Gross domestic product: The total value of goods and services produced in a country is important to development. If there are too many babies then the government will have to spend more on this non-productive sector, and fewer people are able to produce more goods and services. Birth rates tend to be highest in Jamaica among the rural poor, and only slightly lower among women who live in the inner city.

Migration

There is more emigration (people leaving the country permanently) than immigration (people entering the country and remaining permanently) in Jamaica. The annual population increase would be even higher if people did not emigrate. Jamaica's population increase is more a result of natural increase than of migration.

▲ *Table 10.2 Population indicators for Jamaica*

The factors affecting population growth in developing countries are slightly different from those affecting the Caribbean countries. However, some Caribbean countries, for example Barbados, have made important strides towards development and have similar population characteristics to those of a developing country.

Factors affecting population growth in the United States

▲ *Figure 10.9 The USA has a high standard of living*

▲ *Figure 10.10 Many women in the USA hold positions of power*

➤ *Table 10.3 Population indicators for the USA*

Social

Increased affluence and a high standard of living: These factors have lowered infant death rates, and death rates in general. This in turn has increased life expectancy – people in the USA live longer.

Better nutrition, less disease: US society is very food and nutrition conscious, as food technology is a large part of the modern lifestyle.

Women empowered to choose careers over childbirth: There are more educational opportunities and jobs open to women. For example, two women have been appointed to the powerful position of Secretary of State. Many other women hold other government posts.

Later average age for marriage: More women marry later, although there is also a high rate of teenage pregnancy.

High cost of raising children: The high cost of raising and educating children is a deterrent to having a large family.

Availability of birth control and abortions: Despite controversy over the issue of abortion, it is still a strong factor in the lower US birth rate. Contraceptives are readily available.

Economic

Second-highest GDP in the world.

Migration

Immigration is a significant factor in the increase of the US population. This is not just because of the number of arrivals; migrants generally have higher fertility rates than the non-migrant population. Also, in the late twentieth century the government relaxed the controls on immigration, so there was a general rise in the immigrant population (this has changed since terrorist attacks on 9 September 2001).

Go to: www.prb.org and www.statinja.com for more information on US and Jamaican population and migration.

Table 10.4 Population growth in Jamaica and the USA over the past 20 years

Year	Population (000s)	
	Jamaica	**USA**
1995	2483	263119
1996	2510	265284
1997	2534	267636
1998	2557	270299
1999	2574	275562
2000	2589	281421
2001	2605	281423
2002	2618	285094
2003	2630	287974
2004	2645	293655

Exercise

1 Using Table 10.4, create a comparative bar graph of the populations of Jamaica and the USA.

2 Describe the similarities and differences in the birth, death and migration rates.

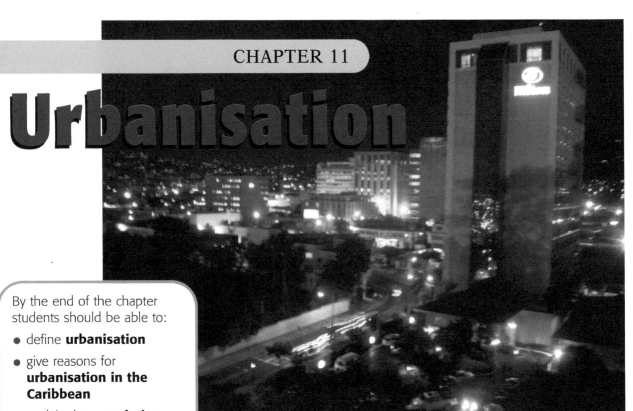

CHAPTER 11

Urbanisation

By the end of the chapter students should be able to:

- define **urbanisation**
- give reasons for **urbanisation in the Caribbean**
- explain the **population growth** within the past 20 years of one capital city in a Caribbean country
- describe the **benefits** and **problems** of urbanisation in that Caribbean capital
- describe ways in which urbanisation can be **controlled** in the Caribbean
- describe the **pattern** and **consequences** of **international migration** in one named Caribbean country within the past 20 years.

✳ **Urbanisation** is the process whereby an increasing percentage of the population of a territory lives in the towns and cities. Usually these people have moved from the rural areas into the towns – **rural to urban migration**.

Reasons for urbanisation in the Caribbean

Urbanisation is one of the most significant human processes affecting some Caribbean societies.

The movement of people from rural to urban areas in the Caribbean increased after the abolition of slavery, and continued after the switch from plantation-based societies to manufacturing centres in the twentieth century. Capital cities were the centres of business and commerce and the only places where jobs other than farming could be found. As rural areas declined in terms of activity and appearance, people were pushed out towards the cities that pulled them in through the many opportunities they offered. Geographers call the factors that influence these movements push and pull factors.

➤ *Figure 11.1 Push and pull factors*

PUSH   PULL

PEOPLE

Push factors

Several factors may drive people out of a rural area.

- **Landless peasants:** large families may outgrow the small plot of land that their parents and grandparents farmed and so be unable to continue their livelihood as farmers. Many farming families in the hills of Jamaica, St Vincent, St Lucia and Dominica have lost a large number of family members to the cities as they leave in search of jobs.

- **Failure in the agricultural sector:** the removal of subsidies by the European Union from banana and other local crops has dealt a devastating blow to many banana farms, especially those of the Windward Islands in the eastern Caribbean. Farming was the main economic activity for many rural folk and they have been forced to leave their farms in search of opportunities in the city.

- **Land degradation:** drought, infertile soils and soil erosion have contributed to the abandonment of agricultural land. Small farmers in some Caribbean territories had already been relegated to the less fertile land, sometimes on steep slopes. These plots are easily degraded after just a few years of planting. Haiti today is experiencing dramatic soil erosion resulting from deforestation in order to create more farmland, and fuel wood harvesting.

➤ *Figure 11.2 An abandoned farming community*

- **Mechanisation of farms:** in Barbados and Jamaica, farmers now use combine harvesters on the sugar plantations. This reduces the number of cane cutters needed during harvest time. Sugar cane workers, both old and young, have had to look for an alternative livelihood in nearby towns.

- **Lack of services:** people may also be pushed out of rural areas because of the lack of social services such as schools, hospitals and recreational facilities. Young people in particular tend to leave rural areas in search of a better life in the city.

Pull factors

Urban areas in the Caribbean are attractive to people.

- **Employment:** industrial and commercial activities in the cities offer more employment opportunities. These used to be located in the capital cities where development was concentrated. However, over the past decade tourism has contributed to the development of many rural towns, offering lucrative employment opportunities. For example, the towns along the south and west coasts of Barbados have grown as a result of tourism. Industrial estates such as the Free Zone in Jamaica and Point Lisas, Trinidad and Tobago, also provide areas of concentrated employment opportunity. Jobs in the city not only offer higher salaries but are also more diverse. There is also a better chance of finding a second job in the city.

- **Social services:** cities and towns in the Caribbean have better social services than rural areas.
  - **Educational institutions (secondary and tertiary)** are better in the cities. Kingston has two of the largest university campuses in the Caribbean and more than ten other colleges. The university, teachers' college and community college campuses in Barbados are located near Bridgetown. The only tertiary institution in St Lucia is located in Castries.
  - **Hospitals** are located in the larger towns. Many Caribbean countries have limited specialist health resources. Those specialised services that do exist are located in the capital city.

- **Entertainment:** the capital cities are usually the centre of entertainment activity. The 'bright lights/big city' concept has lured many young people to Caribbean capital cities. Bridgetown, Kingston and Port-of-Spain are three of the 'party' capitals of the Caribbean. With the growth of tourism, other towns have recreational activities but these are usually aimed at tourists.

- **Cost of goods and services:** prices are generally lower in the cities. The larger shopping malls, wholesalers and large supermarkets are in the city centres.

▲ *Figure 11.3 Shops along Bridge Street in Bridgetown, Barbados*

However, some Caribbean cities are experiencing a decline in the number of people moving into the city. While people still visit the city for all the activities described above, many are moving to live in nearby rural areas. Counter urbanisation is taking place. In Jamaica, St Catherine has had an increase in population, while that in Kingston is decreasing. Over the past decade many new residential developments have been created in parts of rural St Catherine, and Portmore has expanded. The people presently in these communities moved there from Kingston but still commute to and from the capital for work.

➤ *Table 11.1 The population of Caribbean cities*

Country	Capital and largest city	Population of capital
Antigua and Barbuda	St John's	23 500
Barbados	Bridgetown	92 000
Dominica	Roseau	20 000
Grenada	St George's	4 300
Jamaica	Kingston (KMA)	1 228 200
Puerto Rico	San Juan	433 412
St Lucia	Castries	60 300
Trinidad and Tobago	Port-of-Spain	309 100

Source: Borgna Brunner, *Time Almanac 2004*, Pearson Education, 2004

Population growth of Kingston over the past 20 years

Kingston is the largest settlement in Jamaica, and has been since the end of the nineteenth century. The city of Kingston has been combined with the urban area of St Andrew to create the Kingston and St Andrew Metropolitan Area, now referred to as the Kingston Metropolitan Area (KMA). The population of Kingston has grown rapidly over the years. The main cause of this growth is rural to urban migration. As people moved into the city they had children, so natural increase has also been a major contributor to its growth. Kingston has grown over the decades for several other reasons, too.

● It is the chief administrative centre and seat of government, including all the ministerial head offices.

● It is the largest industrial centre: there is a large, diverse range of industrial activities throughout the city. Marcus Garvey Industrial Zone is the largest area.

● Kingston is a centre for sports and entertainment: Jamaica's largest sporting venues are at Sabina Park and Stadium Park. These are large enough to host international events. There are also several nightclubs, game halls and cinemas in the city.

➤ *Figure 11.4 National Stadium in Kingston, Jamaica*

● It is a centre for education: Kingston has more primary, secondary and tertiary educational institutions than any other town in Jamaica. The University of the West Indies and the University of Technology are regional institutions located in Kingston.

● Kingston is a centre of business and commerce: financial institutions, such as banks, insurance companies and the stock exchange, shopping malls and an array of other commercial activities are located in the city.

The population of the KMA experienced its greatest increase in growth (23.9 per cent) between 1970 and 1981. Since then the rate of growth has declined. In the period 1982–90 the city's population grew by 9.7 per cent, and by 1991–2001 the percentage growth was only 0.25 per cent (see Table 11.3). Over the past 20 years the population of Kingston has declined. This is partly as a result of counter urbanisation. The growth experienced in the 1970s and 1980s contributed to over-population of the city.

Benefits and problems of urbanisation in Kingston

▼ Table 11.2 Kingston, Jamaica: its benefits and problems

Kingston can be described as being 'halfway' in terms of development. The city centre faces all the problems typical of many developing countries, but it also shows improvements that are equal to those in some developing countries.

Benefits	Problems
Employment: • better-paying jobs • more attractive jobs • diverse job market	High unemployment: • beggars in the streets and malls • growth of the informal sector • idle young people
Better housing and schools: • universities and colleges • learning centres • private schools • specialist schools • wealthy homes • apartment buildings	Development of ghettos and slums: • poor conditions in schools • poor housing conditions • noise pollution • disease and insanitary conditions • solid waste
Entertainment and recreation: • National Stadium • international nightclubs • theatres • sports complexes	Social decay – crime and violence: • gang warfare in the inner city • drugs and guns • political garrisons
Commerce and industry: • diversity of shopping malls • specialised services, e.g. day spas • fashion houses • industrial estates	Informal economy: • proliferation of peddlers and street traders • illegal drug trade • illegal goods
Telecommunications technology: • all urban centres in Jamaica have access to phones • most government ministries are online • computer cafés in city • multiple cell phone providers	Technology-related crime: • rise in cell phone theft • crimes associated with cell phones and other technologies
Transportation: • bus service • taxi companies • improved road service (roads widened) • internal air service connecting other towns • international airport • large port/wharf	Traffic congestion: • too many cars • air pollution • long hours in traffic • illegal taxis

The high level of unemployment, especially among young people, many of whom are unskilled, has increased poverty in Kingston. People come in from rural areas but are unable to find good jobs and end up in slums and squatter settlements. This leads to other social ills and a vicious cycle of poverty begins. Poverty is the main cause of social problems. As counter urbanisation takes place in other towns, they too are experiencing similar benefits and problems to those in Kingston.

Exercise

Use the information in Table 11.3 to answer the following questions.

1 a Which town had the lowest average annual rate of growth between 1991 and 2001?

   b Is this town a parish capital?

   c Which town had the highest average annual rate of growth between 1991 and 2001?

   d Is this a parish capital?

   e Which parish has the greatest urban conglomeration?

2 Suggest reasons for your choice in 1e (you may need to use a map of Jamaica to help with your answer).

3 Define *counter urbanisation*. To what extent do you think this process could be occurring in the capital city of Jamaica?

▼ *Table 11.3 Jamaica: population in parish capitals and other selected areas, 1991 and 2001*

Parishes and important towns	Population 1991	Population 2001	Average annual growth rate (%), 1991–2001
Kingston and St Andrew Kingston+ and St Andrew Metropolitan Area	563 515	577 623	0.25
St Thomas Morant Bay+	9 602	10 746	1.13
Portland Port Antonio+	13 549	14 541	0.75
St Mary Port Maria+	7 196	7 707	0.69
St Ann St Ann's Bay+	11 051	10 506	−0.50
Ocho Rios	10 254	15 714	4.36
Trelawny Falmouth+	7 955	8 169	0.27
Duncans	1 849	2 125	1.40
Clarks Town	3 139	3 936	2.29
St James Montego Bay+	82 228	95 940	1.55
Cambridge	3 384	3 971	1.61
Anchovy	3 633	4 053	1.10
Hanover Lucea+	5 902	6 036	0.22
Hopewell	4 268	4 739	1.05
Westmoreland Savanna-la-mar+	16 340	19 809	1.94
Negril*	4 184	5 823	3.36
Little London	3 596	4 614	2.52
St Elizabeth Black River+	4 177	4 124	−0.13
Santa Cruz	10 511	10 769	0.24
Junction	3 427	3 625	0.56
Southfield	3 048	3 405	1.11
Manchester Mandeville+	39 946	47 886	1.83
Porus	5 095	6 549	2.54
Williamsfield	3 374	4 237	2.30
Clarendon May Pen	47 700	57 385	1.87
Hayes	8 447	10 062	1.76
Chapleton	3 930	4 544	1.46
St Catherine Spanish Town+	111 201	131 056	1.66
Portmore and Hellshire	96 225	159 974	5.21
Old Harbour	12 718	23 610	6.38
Linstead	9 433	15 046	4.78
Ewarton	6 534	10 699	5.05
Bog Walk	6 572	10 735	5.03

+ *Parish capital towns*
* *Includes the count for the section of Negril located in Hanover*
Source: Statistical Institute of Jamaica: Census 200

Controlling urbanisation in the Caribbean

Some Caribbean countries have made efforts to control urbanisation. For example:

- **Development of new towns:** Portmore in Jamaica was developed primarily to solve the housing shortage in Kingston. It is now one of the largest residential developments in the Caribbean and has recently been awarded municipal status by the government.

- **Decentralising development:** prior to the 1980s, all commercial and industrial activities were focused in the capital cities. This caused rural to urban migration, which led to over-population in these cities. The development of tourism in some smaller towns pulled away some of the urban population. Montego Bay and Ocho Rios in Jamaica are now large urban centres. The construction of Jalousie Resort in Soufrière, St Lucia, has provided many jobs both for people from local communities and for others outside the area. It has also created a **multiplier effect** for other tourism-related activities.

➤ *Figure 11.5 Ocho Rios, Jamaica: an urban tourist centre*

- **Diversifying agriculture:** plantation agriculture and traditional crops have failed in the region in the past two decades. While some territories have shifted the economic focus from agriculture, others have diversified the crops grown in order to maintain the rural economy. Non-traditional exports such as pepper, citrus fruits, spices and vegetables are now being grown for export.

- **Sustainable development:** 'heritage tourism' and 'ecotourism' are being promoted as a means of conserving rural environments while providing jobs for people. Eight Caribbean territories have adopted the UNESCO Youth PATH (Poverty Alleviation through Heritage Tourism) programme as a means of providing a livelihood for young people in rural areas while encouraging them to stay and develop these areas. Bahamas, Barbados, Belize, Jamaica, St Lucia, St Vincent, Surinam and Trinidad are all participants. Heritage and ecotourism are also primary focuses for of the Caribbean Tourism Organisation (CTO) for sustainable tourism in the Caribbean.

➤ *Figure 11.6 Holywell, Jamaica: young people in rural communities being trained as tour guides*

Patterns and consequences of international migration in the Caribbean in the past 20 years

✳ **Migration** is the movement of people from their home in one place to another place. It may be a temporary or permanent move.

Permanent international migration is the movement of people between countries. Within the Caribbean there is **regional migration** (between Caribbean countries) and **international migration** (to other continents such as North America and Europe). Most Caribbean residents migrate to the United States of America (USA) and Canada in North America, or to the United Kingdom (UK) in Europe. These countries have the same, if not greater, attractions as those described on page 158.

Most of the migration in the Caribbean is **voluntary** as people choose to leave for perceived opportunities. However, there are times when migration is **involuntary** (Table 11.4).

▼ *Table 11.4 Reasons for migration and the resulting pattern in the Caribbean*

Voluntary	Examples	Involuntary	Examples
Employment	In every Caribbean country a percentage of its labour force has migrated to the USA, Canada and the UK to work on advertised work programmes in areas such as farming, teaching, nursing, cruise ships etc. Salaries in developed countries are usually higher.	Political	Changes in government and civil unrest have displaced people whose lives are threatened. 750 Haitian refugees landedon Jamaica's shores in 2004 because of political unrest in Haithi. Many tried to reach the USA. Cubans have fled from the communist system of government over the past 30 years.
Study	Migration takes place regionally, especially between UWI campuses. Barbados, Jamaica and Trinidad have centres that prepare students for universities in the USA, and many of these students earn scholarships to go to the USA.	Natural disaster	Hurricanes, earthquakes and volcanoes have displaced people in the Caribbean. Hurricane Ivan forced many people out of the Cayman Islands in 2004. The eruption of the Soufrière Hills in 1995 (and later) forced many people out of Montserrat. Many went to the UK and USA.
Personal	New York, London and Toronto attract many Caribbean people who just want a better life. Living and working in these cities can improve their personal image.		

➤ *Figure 11.7 Haiti: migration as a result of civil unrest*

Haiti – pattern and consequences of migration

Economically and politically, Haiti is a very unstable country. Several governments have been overthrown, the most recent being that of President Bertrand Aristide, in February 2004. This and other factors have contributed to Haitians leaving their country (or *migrating*).

- **Religious and political persecution:** many of Aristide's supporters had to flee after the coup. During the US-led invasion in 1991–93, over 40 000 people fled to the USA as refugees (though not all were accepted there). Over 700 people entered Jamaica from Haiti in 2004 and applied for political asylum. Between 2004 and 2005, the number of 'boat people' arriving on other Caribbean islands also increased.

- **Civil war:** each government coup is accompanied by civil unrest, so people want to leave the country.

- **Famine:** many Haitians are experiencing a lack of food due to environmental degradation and economic problems. Over 50 per cent of the population lives in poverty. Every year the Dominican Republic receives a large number of refugees from Haiti.

- **Natural disaster:** Hurricane Jean caused over 2 000 deaths in September 2004. Several communities were destroyed or damaged by landslides and floodwaters. People were without food for weeks. Many tried to flee to other Caribbean countries and to Florida in the USA.

- **Over-population:** Haiti has a population of approximately 8.1 million, about 38 per cent of whom live in the capital, Port-au-Prince. Haiti's population has outstripped its resources, forcing people to leave the country.

- **Chronic poverty:** social services such as health, education, utilities and housing are all poor in Haiti. Just over 50 per cent of the population has access to safe water and sanitation services. More than 50 per cent are illiterate and only 43 per cent are economically active. Those who can escape do so. Usually, however, it is those who have some education and access to money who migrate. Most people try to accumulate savings and leave as soon as they have the opportunity.

The consequences of migration from Haiti can be devastating for the country. However, in some cases it can be positive for both the migrant and the government.

- **Remittance:** those who migrate usually improve their personal status and send money home for those left behind. The government indirectly earns an income from the tax on remittances, when people spend this money. It also shifts some of the burden from the government to provide welfare for these beneficiaries.

- **Population change:** migration reduces the population pressure on resources in Haiti. Those who are left behind have a better chance of employment as there is less competition for scarce jobs.

- **Political freedom:** migrants are able to live in peace without fear for their life.

- **Brain drain:** one of the greatest negative impacts of out-migration on developing countries is the drain on the skilled sector. The host country may invite educated and skilled people, such as farm workers and teachers, to help build the economy of that country, while the home country is left with a further shortfall in its human resources.

- **Deportees:** a new social problem affecting some Caribbean countries is the return of 'deportees'. These are people who have been sent back to their home countries after committing crimes (sometimes serious) in the countries to which they migrated. Countries like Haiti are ill-equipped to deal with such people.

However, the overall consequence of migration from Haiti is the inability of the country to maintain its own population at a satisfactory standard of living, which would enable it to build a strong economy and society.

Exercise

1 Use the Internet to collect statistics showing the pattern of migration in a Caribbean territory of your choice.

a Create a table showing out-migration and the countries where most migrants move to.

b List some of the effects of international migration on your selected country.

2 Organise a class debate on the motion 'This house believes deportees are a burden on Caribbean societies.' How might the issues discussed be resolved?

CHAPTER 12

Primary economic activities

By the end of the chapter students should be able to:

- describe the characteristics and relative importance of **the different types of economic activities** in the Caribbean

- locate one example of a primary activity – **fishing**, **forestry** or **mining**

- explain **factors influencing the location of** that primary economic activity

- describe the **trends** and explain the **challenges** of your chosen example

- describe the **importance of agriculture** to the Caribbean region

- explain the **changing role of agriculture** in the Caribbean economies

- locate areas in at least *one* Caribbean country where **commercial arable and peasant farming** are important

- describe and compare the **characteristics of commercial arable and peasant farming** in that country with **wheat farming in the Prairies of Canada**

- compare the **trends in commercial arable farming** in that country with wheat farming in the Prairies of Canada.

✳ **Economic activities** refer to any action that is carried out to produce goods and services in exchange for currency. At the end of the activity, trade takes place and income and value are added.

✳ **Primary economic activities** are those involved in using natural resources, such as the extraction of minerals from the Earth, fishing from the sea, forestry and agriculture from the land. They involve no change in the natural resource, and no processing takes place.

✳ **Secondary economic activities** involve the processing and altering of raw materials (or components of raw materials) to produce new products. The term commonly used is *manufacturing*. Materials are transformed from one state to another to satisfy different requirements.

✳ **Tertiary economic activities** are those that provide services rather than a tangible product. These include retailing, banking, transportation and tourism industries.

Relative importance of different types of economic activity

Traditionally, the Caribbean colonial economies were dominated by primary economic activity such as agriculture and mining: essentially producing raw materials for the secondary economic industries of the 'mother countries'. The colonies were in fact forbidden by law from developing any secondary industries beyond basic sugar factories, oil refineries and bauxite drying.

Post-independence, many countries instituted 'industrialisation by invitation' schemes, which included the development of industrial estates/parks and tax/duty-free incentives. Secondary industries (manufacturing) became more important to many economies. Construction activities were significant employers, as were machine operators. There was a trend away from the traditional primary activities and towards fulfilling the expectations of the workforce for non-agricultural employment.

In the last two decades, many manufacturing industries have closed and gone to cheaper locations, especially in the Far East, with only Trinidad and Tobago retaining and expanding their manufacturing sector, based on abundant cheap natural gas supplies. (Government policy created the Point Lisas Industrial Park for the only iron and steel smelter in the English-speaking Caribbean; as well as more recently the methanol industry supplying 70% of USA demand.) Most other territories retain agro-based and consumer-oriented small manufacturing sectors or export-oriented sectors such as the Free Zone of Kingston, Jamaica.

At the start of the twenty-first century, with falling prices for agricultural products and international competition from other locations, many Caribbean countries have turned to tertiary economic activities, i.e. services such as tourism, information technology and financial services to satisfy the demand of its educated population for more skilled white-collar jobs. Tourism is a major employer on many islands and makes a significant contribution to the economies of The Bahamas, Barbados, Jamaica, St Lucia among others.

Many economies have experienced a dwindling agricultural sector (St Kitts deciding to abandon sugar cane all together after 2006), a modest manufacturing sector and an expanding tertiary sector.

The following are some reasons why economic activity is important to the Caribbean.

- **Income and revenue:** economic activities make up the gross domestic product (GDP) of a country. The GDP is the value of all the goods and services that are produced in a country. Basically it is a measure of income. Different economic activities contribute varying amounts. Tourism, for example, contributes 15.5 per cent of the GDP of St Kitts and Nevis.

  GDP is used as an indicator of economic growth and success and can be used to measure the standard of living. The extent of economic activities is therefore important for all Caribbean countries.

➤ *Table 12.1 GDP of some Caribbean territories*

Country	GDP per capita (2003)
Bahamas	$17,700
Guadeloupe	$7,900
Martinique	$14,400
Cayman Islands	$32,300
Turks and Caicos Islands	$11,500
Aruba	$28,000
Netherlands Antilles	$11,400

● **Balance of payments deficit:** when manufactured goods are exported, they reduce the amount of foreign exchange spent on imports while increasing export earnings. This reduces the amount of borrowing and spending of US dollars, therefore reducing the debt burden.

● **Employment:** employment describes the involvement of people in economic activities in exchange for income. The labour force of a country is important to its economic growth. People earn and then they spend. The exchange of money in this way creates a multiplier effect in the economy.

Employment in manufacturing can create jobs for farmers in the primary sector and retailers in the tertiary sector. Employment increases purchasing power (the ability to pay for goods and services). The government also earns taxes from income. Each individual who is employed has to pay a percentage of his/her wages to the government as income tax (some individuals are exempted because they do not earn enough).

The greater the diversity of economic activities, the greater the income and revenue will be. Economic diversity also creates more employment opportunities.

Exercise

Part 1

1   Make a list of the occupation of the parents of everyone in the class. (For someone who is self-employed, enter 'employment'.)

2   Each student should write the occupation of either or both parents on separate pieces of paper. (Do not write your name or your parents' name on it.)

3   Put the strips of paper in a box and shake.

4   Write a large label on the class board:

'Employment in _____' (write the name of the town or country you are in).

Create a three-column table, with headings showing the categories of economic activities:

Primary	Secondary	Tertiary

5   Each student now randomly picks a paper from the box and sticks it (using a thumbtack or appropriate fastener) to the board under the correct category of economic activity.

6   Try to create a web by joining related activities with a line.

7   Name the most common occupation, and the category with the most employment.

8   Discuss how some activities in one category (from one list) may be related to others in the other lists/categories on the board.

9   Make a list on the board of what job each student would like to do in the future.

Part 2

Write an essay (about 500 words) on what life might be like if both your parents were unemployed for an extended period.

Exercise

1 Draw a simple flow chart showing the relationships between one manufacturing industry in your town or country and the other economic activities nearby. The following may be included on the chart:

- shops – retailing
- transport – buses that take people to and from work (other types can be mentioned)
- entertainment – carnival, festivals, stage shows, cinemas/theatres
- farming – market outlets
- education – schools and other centres of learning.

2 Make connections to different types of economic activity (primary, secondary and tertiary).

- **Development of infrastructure:** a percentage of income earned from economic activities goes towards the national income in the form of taxes. Complex webs of taxes are charged at various levels of economic activity. These taxes help to develop the infrastructure: roads, bridges, transportation facilities, etc. Many foreign investors are encouraged to establish themselves in the Caribbean. These investors set up economic activities that provide jobs while the government earns taxes. Some of these investors may actually put in the infrastructure to facilitate their operations. In exchange the country enjoys the benefit of development that these may add.

- **Transport and trade:** economic activities stimulate and increase transportation and trade through the development of land, air and sea routes. Goods and services must be moved from one place to another, so the more economic activities there are in a region, the more movement of goods and people there is.

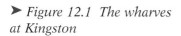
➤ *Figure 12.1 The wharves at Kingston*

Exercise

1 Draw a sketch map of your country and show the following on your map:
   - the major road network
   - all air and sea ports
   - railway lines and stations
   - major industrial areas (manufacturing).

2 Describe the patterns observed between the location of transport routes and the location of factories.

✳ **Resources** are features of the natural environment that are used by people. People's knowledge and skills that are used for economic benefit are referred to as **human resources**.

✳ **Renewable resources** are those natural resources which, if they are properly managed, will regenerate over short periods of time, for example water, soil fertility and trees.

✳ **Non-renewable resources** are those which, once used up, will take thousands or even millions of years to regenerate in the natural environment, for example oil and bauxite.

➤ *Figure 12.2 The resources extracted in primary production are natural – some are renewable and others are non-renewable*

Primary activities are important because the outputs from these activities provide materials for direct use and raw materials for other industries. The primary sector provides goods for export. Many exports from the Caribbean are based on primary extraction of minerals and the products of agriculture.

Location of primary economic activities in the Caribbean

▼ *Figure 12.3 Forestry in Guyana*

▲ *Figure 12.4 Oil and natural gas in Trinidad*

▲ *Figure 12.5 Bauxite mines in Jamaica*

◄ *Figure 12.6 Main fishing ports in Belize*

▼ *Table 12.2 Some primary economic activities in the Caribbean*

Economic activity	Factors influencing location
Bauxite (Jamaica)	• *Raw material:* lies near to surface – easily accessible. Plants built near deposits. • *Transport:* mineral transported by road, rail and conveyor belt. Road and rail facilities are constructed by bauxite companies in mining areas. • *Capital input:* expensive outlay for equipment, which must be maintained. Most bauxite companies in Jamaica are partly owned by foreign investors which provide capital. • *Labour supply:* requires a highly skilled labour force (technical skills, e.g. machine operators). Towns like Mandeville have sprung up around mines. • *Ports and markets:* bauxite is extracted for export. It is bulky and therefore should be near to or have easy access to port facilities. Jamaica's ports are near to mines (see Figure 12.3). Bauxite mined more than 10 km from ports is transported by rail and conveyor belt.
Oil and natural gas (Trinidad)	• *Raw material:* oil and natural gas lie primarily in the southern half of Trinidad. • *Transport:* oil and gas are transported by pipeline to the processing plants. • *Capital input:* drilling is expensive, especially when little or no oil is found. Capital is injected by both local and foreign investors. • *Labour supply:* like bauxite, oil exploration and mining require highly skilled labour. The engineering faculty of UWI is able to train personnel, reducing the need for foreign expertise. • *Processing plants* are near ports.

Forestry (Guyana)	• *Raw material:* Guyana's forests are the largest (84% of the country is covered in forest) in the Caribbean because:
	  • large amount of rainfall (2 000–3 000 mm)
	  • deep tropical soils
	  • high temperatures
	  • several types of forests and a variety of tree species.

Fishing (Belize)	• *Raw material:* fish are abundant along the coast of Belize because of the extensive reef (barrier reef) and mangrove cover (excellent breeding ground for fish) along 50+ km stretch of coastline. Products include fish, lobster, shrimp, conch. • *Market:* fish in Belize, as in the rest of the Caribbean, is sold along the shoreline. Buyers come to the shore as fishermen come in. Some of the buyers are wholesalers who retail the products to supermarkets and hotels. Others have arrangements for special catches of shellfish to be exported. • Most *fishermen* live along the coastline, making it easy for them to carry out fishing as an economic activity.

Economic activity	Trends in production
Bauxite (Jamaica)	Production has fluctuated over the years primarily because of global trends in demand for bauxite. • World price has fallen as alternatives for aluminum are explored. Plastic is a popular alternative. • Some countries stockpiled the product when prices were favourable, so demand has fallen. • Production is also affected by issues with the overseas parent company. For example, if the parent company is having economic problems then production will be affected in Jamaica. • In recent years there has been a move to use more local expertise, so there is less need for overseas technical assistance
Oil and natural gas (Trinidad)	• Crude oil production has declined over the years. However, the economy of Trinidad has diversified primarily because of oil and natural gas production. Natural gas provides the energy for other industries, and petrochemicals are produced from oil. • The war on terrorism (in particular the war in Iraq) has affected world prices, pushing them upwards. Trinidad is in an advantageous position. • Refineries have upgraded over the years to improve quality of output.
Forestry (Guyana)	Several measures have been taken to make forestry a viable economic activity. • Exotic species have been introduced which grow faster over large areas and are easier to harvest. Examples include pine and teak. • New roads have been built to give access to forested areas. • Conservation techniques are now being applied. Reforestation is encouraged to replace trees that have been commercially removed. • New regulations have been set up to monitor and ensure effective use of forest resources. • New markets have been identified for hardwood trees. • Forest is being used for other economic activities such as ecotourism (see Chapter 13). Ecotourism is also being used as a means of conservation of forest resources.
Fishing (Belize)	• The trend has been an organisation of the fishing industry especially to meet export demands. Fishermen are now joining co-operatives in order to enjoy benefits such as increased political representation. • Keeping up with global trends in processing is still a challenge. • Conservation techniques are now being applied in order to sustain the industry and to save marine habitats (coral reefs and mangrove swamps). • The industry is now linked to tourism – production of fish and ecotourism (tours of the reefs).

Economic activity	Challenges faced by the industry
Bauxite (Jamaica)	• Fluctuation in world prices affects profitability. When prices fall profit margins fall, affecting stability of industry (workers may have to be laid off). • Alternative sources of bauxite and alternative products such as plastic are affecting the demand for aluminum. • Environmental degradation occurs as a result of pollution during mining. Liquid and gaseous chemicals are expelled into the air and ground. Residents living near plants complain of foul smells, dust and loud noise. Mining also detracts from the aesthetic beauty of the environment. Companies are faced with the (costly) challenge of correcting or reducing pollution.
Oil and natural gas (Trinidad)	• Exploration may be haphazard, despite the use of high-tech equipment. • Fluctuation in grade (quality) of oil makes refining more costly. • Oil prices must be in line with OPEC (Organisation of Petroleum Exporting Countries), so prices are set. • Oil reserves are limited. • Environmental degradation occurs through seepage and accidents. Coastal waters are especially vulnerable. Oil spills kill marine life and take a long time to clean up. Clean-up is also expensive. The underground water supply is sometimes contaminated during drilling.
Forestry (Guyana)	• Extraction of particular trees is difficult because there are many varieties of trees in the forest. • Inaccessibility is a major problem as Guyana's road network is limited. Rivers are used to transport logs, but some are lost in the rainy season and in dry season the water level may be too low for transporting them. • Demand for hardwood is much lower than for softwood. Other resources such as concrete have replaced the need for hardwood. Softwood is also easier to transport.
Fishing (Belize)	• Fishing grounds have been destroyed due to bad fishing practices. • Ill-equipped boats and lack of proper storage facilities mean that when there is difficulty at sea the catch is more than likely to spoil. • Overfishing: despite the abundance of fish in Belize's waters, fishermen tend to catch some varieties out of season. • Pollution from land-based activities, such as sewage and sedimentation from soil erosion, threaten coral reefs and the creatures that live there.

Exercise

1 a Identify two, non-agricultural, primary activities in your territory.

  b Why have you selected those two?

2 Identify a natural resource you think could be developed in your country. Write a letter to the Prime Minister outlining your ideas for development.

The importance of agriculture

Agriculture differs in some very important respects from other primary activities such as mining and forestry. The cultivation of the soil is a widespread and important economic activity. It is through the cultivation of the soil that we eat. Most of the foods we eat are cultivated (crops) or reared (animals) on farms, although it may not appear that way – the natural state of some ingredients can be changed beyond recognition by the manufacturing process.

Exercise

Wheat and cane sugar are the basic ingredients in many processed foods. Make a table and list as many foods as you can that have:

- wheat as an ingredient
- cane sugar as an ingredient
- both ingredients.

▼ *Figure 12.7 Food crops on display in a market*

✳ **Agriculture** is the cultivation of crops and the rearing of animals for food and non-food resources.

Agriculture is one of the most basic economic activities and still accounts for a large part of human activities in the Caribbean.

'No country can afford to put farming on the back burner because no country can be self-sufficient unless it has the capacity to feed itself.'

Roger Clark, Minister of Agriculture, Jamaica – November 2003

In 2002, agriculture contributed US$22.8 billion to Jamaica's Gross Domestic Product (GDP).

➤ *Figure 12.8 Contribution of agriculture to Caribbean economies*

Country	Percentage of GDP resulting from agricultural revenue
Antigua and Barbuda	3.9
Barbados	6
Dominica	18
Jamaica	6.1
St Lucia	7
Trinidad and Tobago	2.7

Agriculture is important to the Caribbean in other ways, too.

- **Employment:** many rural families within the Caribbean region are employed in agriculture.

➤ *Table 12.3 Agricultural labour force in selected Caribbean countries*

Country	Percentage of workforce employed in the agricultural sector
Antigua and Barbuda	7
Barbados	10
Dominica	18
Jamaica	20
St Lucia	22
Trinidad and Tobago	10

- **Food security:** food crops grown on small farms account for a large percentage of food consumption, especially for rural people. Agriculture makes food available and accessible and provides the nutrition needed by people.

- **Economy:** agriculture contributes to the GDP of most Caribbean economies (Figure 12.8). Agricultural products are also the main exports for some Caribbean territories, earning valuable foreign exchange.

- **Linkage industries:** agriculture provides raw materials for food processing and some non-food products like pharmaceuticals, animal foods, fuel (e.g. sugar cane trash) and crafts (e.g. coconut shells and husks are used to make trinkets and jewellery). Agriculture is now a direct link to tourism in many territories as local foods are displayed and served as part of the region's marketing strategy.

- **Heritage:** much of the culture and tradition of indigenous Caribbean people had and still has its roots in farming. For example, the Caribs of Dominica and the Mayas in Belize built their societies around agriculture. Some territories have special celebrations to mark significant agricultural achievements, such as the Trelawny Yam Festival in Jamaica and the celebrated Crop Over in Barbados.

➤ *Figure 12.9 Crop Over in Barbados*

The changing role of agriculture in Caribbean economies

Agriculture (farming) has changed over time. Most Caribbean territories started out with plantation agriculture as their economic base, around which the society developed. Markets for crops like sugar and bananas were often in far-distant places. European countries that originally colonised Caribbean territories continued to purchase their products long after these countries gained independence. This provided a sense of security for the large farmers and governments of the Caribbean. However, much has now changed both in the processes of agriculture and in the role the sector plays in Caribbean development.

- Agriculture has an essential part to play in a country's development. It is significant in social and economic development as well as in environmental sustainability (see Chapter 15). Farmers are now diversifying and producing for niche (specialist) markets, for example production of flowers for overseas markets, pondfish for the retail industry and specific crops for tourism (hotels). However, this is sometimes to the detriment of the local market as production is aimed only at the niche market and not at local consumers.

- Government policies within the region affect the role agriculture plays in regional development. 'A bad policy can make the most fertile soil infertile,' says the IICA (Inter-American Institute for Co-operation in Agriculture). The input of governments and regional bodies in the development of agriculture will influence the value earned. Policies must make radical changes in production and marketing.

● People must eat! However, what they eat, how they eat and where they eat have changed dramatically over the past two decades. Agriculture in the region must reflect this. It must also capitalise on the changes required for exports, and reduce the need for imported food. This will mean a diversification in crop output from traditional crops.

Some observable trends that can affect the agricultural sector are:
  ● health consciousness – low carbohydrate diets, vegetarian diets
  ● convenience fast foods/ready to go foods
  ● food packaging and presentation
  ● organic food
  ● high-income health-driven consumers versus low-income budget buyers.

These trends will affect marketing and product output. Agriculture has a choice: it can either address these changes, or ignore them; it can move towards fundamental changes and benefits – or lose out.

➤ *Figure 12.10 Local restaurants rely on local agricultural produce*

● Globalisation has played a significant role in stimulating change in Caribbean agriculture. It has:
  ● influenced competition and trade patterns in the world market – Caribbean countries are no longer enjoying preferential treatment from their former markets in Europe
  ● influenced changes in taste patterns – including a taste for foreign foods
  ● exposed Caribbean culture to new places in the world and so has opened up opportunities for new markets – the region's agricultural sector must take advantage of this
  ● influenced the use of technology in the agricultural sector.

➤ *Figure 12.11 Caribbean food on sale in a London market*

If the agricultural/food-product sector is to continue as the foundation of Caribbean society, it must have more influence on the region's economy. This is the greatest challenge for agriculture. Employment in the sector has declined over the last decade. Urbanisation (see Chapter 11) has contributed to the decline as people continue to move from rural areas into cities. Urbanisation has also contributed to a decrease in the land area available for crop production, as good agricultural land is being used for housing.

However, other factors such as economic diversification have also resulted in a shift in the labour force. The oil boom in Trinidad and tourism in Tobago have contributed to the decline in the agricultural labour force in those islands. In 1999, agriculture contributed only 2.2 per cent to the GDP of Trinidad, a decline from the previous years. Tourism development in many Caribbean territories provides alternative employment for rural residents. The agricultural labour force in regions like St Mary, Jamaica, has dropped by a half, as there is more and more demand for workers in the hotels in Ocho Rios.

Caribbean territories have been taking measures to improve and increase the capacity of the agricultural sector. In Barbados a special agro-tourism programme has been introduced through the IICA. This programme aims to increase the potential earning of agriculture by creating non-traditional markets. In Jamaica, the 'Buy Jamaican' campaign is an effort to revitalise the agricultural sector and increase local market capacity.

Go to www.IICA.com for more information on the changing role of agriculture in the region.

Exercise

Conduct a survey of a local restaurant in your country. You will need a copy of the menu.

1 Create a table with four columns.

 a In the first column make a list of all the ingredients on the menu.
 b In the second column put a tick beside each ingredient that can be supplied by local farmers.
 c In the third column put a tick beside each ingredient that can only be supplied from outside your country.
 d In the fourth column list the ingredients the restaurant has to import (directly or indirectly – the sales company may have imported them). If you are not sure, check the packaging.

Example: Soso Restaurant – Jamaica

Ingredients	Locally available/ can be produced locally	Must be imported	Imported ingredients
Rice		✓	Imported
Chicken	✓		Imported
Tomato sauce	✓		Imported

2 Suggest reasons for importing some items that are available locally or that can be made locally.

3 Hold a class discussion on the issue of diversifying agriculture to meet global challenges.

4 In your discussion, decide whether you believe agriculture is attractive for young people to take up as a career.

Different types of farming

✳ **Commercial arable farming** is the extensive cultivation of (usually) one crop (monoculture) for sale.

✳ **Peasant farming** is the small-scale cultivation of crops and rearing of animals for local consumption. The crops may be sold at the farm or in local markets; some of the produce is consumed on the farm.

Peasant farming

Peasant farming is an important feature of Caribbean land use. When slavery ended, many former slaves went to the hills and set up small farms. Some had previous experience with market garden plots on the estates. Food was grown for the family and any excess sold in the local markets. Today, peasant farms are a dominant feature of the Caribbean landscape, and they make a significant contribution to the cuisine and economies of the region.

Characteristics of peasant farming in Trinidad and Tobago

A history of late settlement meant that Trinidad, like Guyana, had few slaves at emancipation, and landowners therefore imported labour from another English colony: India. These indentured labourers were given either 5 acres of land or return passages to India at the end of tenure. Most labourers stayed and became smallholders. East Indian indentured labourers played a large part in the history of the Trinidad sugar cane industry and became former Prime Minister Panday's main constituents – the sugar workers' trade union.

The largest farms (>400 hectares) and all the estates are now state-owned, but the majority of holdings are less than 6 hectares. In 2003, 10 000 sugar workers accepted the Government's offer of a voluntary separation package including being eligible for residential lands. The government also formed the Sugar Manufacturing Co. to take over from Caroni (1975) Ltd to produce sugar on a much smaller scale. Trinidad has only one operating sugar factory; in 2005 it produced 31 000 tonnes and could not meet its EU quota.

Most peasant farms occupy small plots of land but some are larger, producing crops such as vegetables and seasonings. This type of farming is referred to in some places in Trinidad as market gardening. On the Caroni Plain an old abandoned sugar estate has been divided into a number of plots of land for market gardening. Most of the vegetables for the island are grown there.

- The land is intensively cultivated – that is, maximum use is made of the space available.

- The crops planted grow fast and bring in ready cash. These crops may be ready to harvest within six weeks.

- Crops are grown within easy reach of available markets in the towns and cities.

- Some farms use modern technology, which reduces the need for land space and increases output. For example, one technique used in Trinidad is hydrophonics. The crops are usually grown in a greenhouse, in pipes or plastic channels lined with gravel, sand or husks. They are grown in a chemical solution and no soil is needed for this type of cultivation.

Several other techniques are used in peasant farming, for example:

- **Intercropping and mix-cropping:** several crops, usually tree crops and short-term crops, are grown in between each other. This helps to provide the farmer with a year-round income.

- **Crop rotation:** this is used to reduce soil exhaustion, which can result from planting only one crop. This technique also helps in providing the farmer with an income all year round.

- **Terracing (cutting steps in the hillside):** this technique maximises hillside space while reducing soil erosion. It is used in parts of the Northern Range and in some hilly areas of Tobago.

Some peasant farmers grow crops like cocoa for export. This used to be common in Tobago, but production there has decreased in recent years.

Crops such as citrus fruits are grown for the food processing industry (Chapter 13). Trinidad manufactures the largest quantity of juice concentrates in the Caribbean, so local small farmers have a ready market.

The challenges faced by farmers have forced them to diversify their crops, so they no longer concentrate on growing traditional crops only. Instead they grow different crops at different times depending on local and international demands.

Commercial arable farming

Commercial arable farming in the Caribbean was introduced in the sixteenth century with the establishment of sugar cane plantations and the system of slavery. Caribbean society was born out of plantation agriculture. All available flat land was used for the cultivation of sugar cane, with most estates having factory facilities for the processing of sugar, which was exported to European countries. When the sugar industry faced problems with competition from European sugar beet, commercial arable farming was diversified in some territories. Bananas ('green gold') became the main crop in some territories, and tobacco in Cuba. Today sugar cane is still grown extensively in some territories.

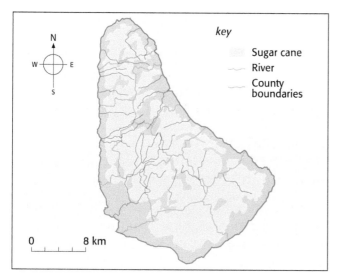

▲ *Figure 12.12 Sugar cane cultivation in Barbados*

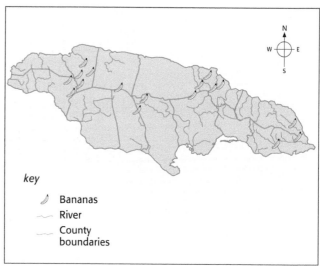

▲ *Figure 12.13 Banana cultivation in Jamaica*

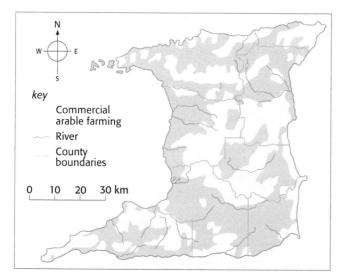

▲ *Figure 12.14 Commercial arable farming in Trinidad*

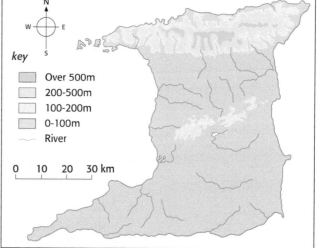

▲ *Figure 12.15 Relief map of Trinidad*

Exercise

Study Figures 12.14 and 12.15.

1 What is the relationship between the two maps?

2 What are the main commercially grown crops?

Characteristics of commercial arable farming in Trinidad and Tobago

The characteristics of plantation agriculture in Trinidad are similar to those in most other territories. Several crops are grown commercially, but sugar cane is the dominant crop grown for export. The main features of commercial arable farming in Trinidad are as follows.

- Large landholdings are on relatively flat land, for easier movement of machinery, for example the Caroni and Naparima Plains. Cultivation in Caroni covers over 31 000 hectares.

- Climate: sugar cane requires an annual average of 1 200–1 500 mm of rainfall. Trinidad receives seasonal rainfall resulting primarily from the influence of the Inter-Tropical Convergence Zone (ITCZ). The Caroni region receives an average of 1 500 mm of rain annually and plenty of sunshine, which are ideal for the cultivation of sugar cane. Because of the mountains to the north and east, farms on the western plains are protected from strong winds.

- Topography: commercial farming takes up most of the available fertile flat land. Sugar cane and other crops are grown commercially along the coastal plains. The west-central region of Trinidad has a greater area of flat land and is mostly under cultivation.

- Monoculture is a dominant feature of commercial arable farming. Sugar cane in Trinidad is a monoculture crop and is the major crop grown commercially. In 2002 the Caroni Estate produced 40 per cent of the sugar cane needed to produce 100 000 tonnes of sugar.

- Export crop: in 1999, sugar cane export made up 33.8 percent of the 2% contribution that agriculture made to Trinidad's GDP.

- A large labour force is required. The sugar cane plantation system requires workers at various levels of specialisation, with a large part of the labour force directly involved in field cultivation. In Trinidad in 1999, 25 per cent of the total agricultural labour force was involved in sugar cane cultivation – a total of 10 400 people.

- Heavy capital investment is required in sugar cane plantations. Following its recent upgrade, the factory at Usine Ste Madeleine in Trinidad now has the most up-to-date equipment and technology. This will raise standards to internationally accepted levels and increase output.

- Machinery such as tractors, ploughs, crawlers and combine harvesters are used on estates to increase production.

➤ *Figure 12.16 High-tech equipment is used on large farms*

- Chemicals are required for the sugar cane crop, such as:
  - pesticides
  - herbicides
  - fungicides
  - fertilisers.

- Problems affecting cultivation include:
  - pests and diseases – smut is a serious threat to sugar cane
  - bad weather – Trinidad is affected by heavy rains from the ITCZ and tropical storms associated with easterly waves
  - fluctuation in world prices and world demand
  - competition from the global market
  - trade liberalisation – the removal of subsidies and preferential treatment presents challenges of unfair competition from larger producers.

Exercise – group work

Create a model layout of a typical sugar plantation.

1 Using neat labels, highlight all the characteristics discussed above.

2 Use your map skills (see Chapter 2) and calculate a scale for your model.

3 Display your model in class.

Characteristics of commercial arable farming in the Prairie Provinces of Canada

➤ *Figure 12.17 The Prairie Provinces of Canada*

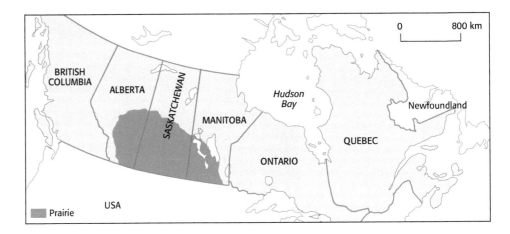

Canada is a developed country where agriculture is a large-scale commercial activity. Wheat is the main crop grown for export. In the early 1900s, the Prairie Provinces of Canada became a major wheat-producing area and later one of the largest exporters of wheat in the world. Wheat cultivation has some specific characteristics which are similar to those listed for sugar cane cultivation.

- Large expanses of flat undulating land: the Prairies are separate from the Great Plains of North America. All the states occupying the plains are engaged in wheat cultivation.

- The climate of the Great Plains is ideal for wheat. The area receives an annual average of 500 mm of rainfall, which is just sufficient for the ripening of the grain. Harvesting is done in summer when it is warmer (an average of 18°C).

➤ *Figure 12.18 Average temperatures and precipitation in Winnipeg*

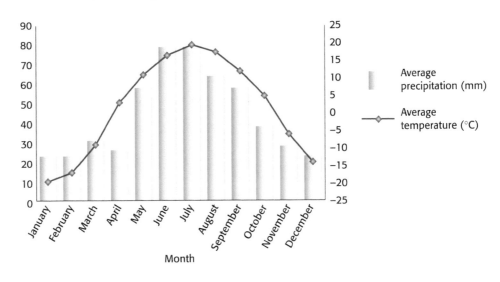

- Export: Canada exports wheat to many parts of the world – the USA, Europe, Asia, South America and the Caribbean.

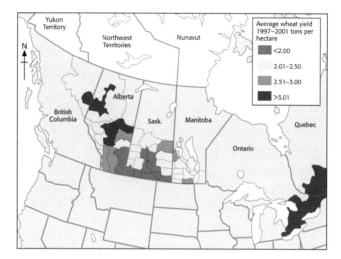

▲ *Figure 12.19 Canadian wheat production*

- Heavy capital investment is a key factor. Most farms today use computerised records and analysis, bio-engineering and other high-technology inputs.

- Machinery and mechanised transport are widely used. There is less dependence on human labour in the fields than in the Caribbean. This is because of the extensive use of mechanised harvesters. The railroad system is highly developed between towns like Winnipeg, Regina, Edmonton, Saskatoon and Calgary, which developed as market and transport centres for wheat.

- There is widespread use of chemicals on crops, such as:
  - pesticides
  - herbicides
  - fungicides
  - fertilisers.

- Problems affecting cultivation include:
  - pests and diseases – locusts and grasshoppers are a threat to the crop; smut also affects wheat
  - bad weather – hailstones during storms, and frost during the long winter, affect the wheat
  - fluctuation in world prices – some years there is overproduction and excess wheat causes prices to fall; the opposite happens when the crop is affected by problems and so is in short supply.

▼ *Table 12.4 Trends in commercial arable farming in the Caribbean and the Prairie Provinces of Canada*

Table 12.4, below, outlines some of the differences and similarities between arable farming in the Caribbean and the Prairie Provinces.

Trends	Caribbean	Prairie Provinces in Canada
Diversification	Reduction in production of traditional crops has led to introduction of new crops such as: ● citrus and other fruits ● vegetables ● ground provisions (root crops) ● spices ● horticultural produce	Other crops grown include: ● barley ● maize ● rapeseed In addition, mixed farming has increased
Globalisation	Opening up of new markets Changing structure of farms to meet global standards – some farms have certification with international certification institutions, e.g. Green Globe (2003)	Increase in demand for wheat as food products diversify and demand for wheat products increases Farms getting certification
Trade agreements	Signing of trade agreements, e.g. WTO, FTAA and CSME, to improve relationships and protect markets	Signing of trade agreements, member of NATO, NAFTA and OECD
Linkage industry	Food processing has increased over the past decade providing markets for local producers	Increase on world market means increase for Canada
New technology	Farms in the Caribbean are using new equipment, including computers	Upgrade of equipment Technology applied to reduce environmental damage

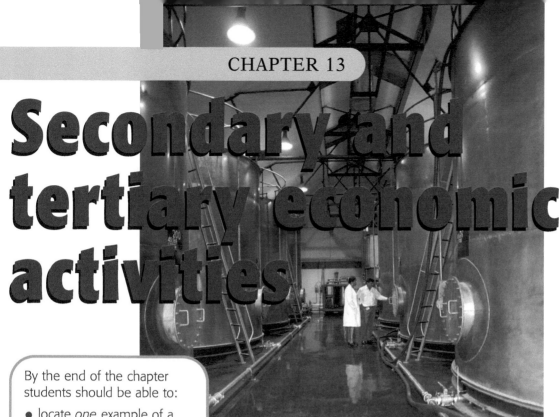

CHAPTER 13

Secondary and tertiary economic activities

By the end of the chapter students should be able to:

- locate *one* example of a secondary economic activity (**garment industry or food industry**) and a tertiary economic activity (**tourism**) in the Caribbean

- explain the **factors influencing the location** of these economic activities

- describe the **trends** and explain the **challenges** faced by these economic activities

- compare the **garment industry** in a named Caribbean territory with that in a **newly industrialised** island state (Singapore).

In Chapter 12 we learnt about agriculture and other primary economic activities. In this chapter we look at secondary activities (manufacturing) which turn raw materials into more useful forms, and at tertiary economic activities, such as tourism, which provide services. Tertiary economic activities are very important in the Caribbean.

Secondary economic activities

The activities of manufacturing take place in large factories (industrial areas) and involve the use of specialised machinery. When production occurs on a small scale, such as in homes or small workshops, it is classified as cottage industry. Manufacturing involves certain features that influence the location and development of the activities.

- **Mechanisation:** manufacturing in factories marked the start of the Industrial Revolution. Large machines serve to improve upon human labour and make it more efficient. In large factories no single individual is responsible for making a product from start to finish. One product may be exposed to 20 sets of workers during the production stage. A shirt, for example, will take one individual 3 hours to make using a single machine. However, in a factory it may take 5 people at 5 stations (using high-technology machines) to make 600 shirts in the same timeframe. (How many shirts does each individual make per hour?) With the division of labour, mechanisation increases the pace and volume of output.

- **Structure and linkage:** manufacturing is not generally the processing of just one raw material and the sale of the product. In many instances the activities involve several stages and processes using the products and services from other industries. The food processing industry, for example, uses raw material from the agricultural sector; packaging from the pulp and paper manufacturer, or plastic (polythene) from the petrochemical industry; biochemical products such as artificial flavours and sweeteners; and nutrients such as vitamins and minerals produced by pharmaceutical companies.

- **Value added:** manufacturing adds value to the raw materials it uses. The finished product is worth more than the total value of the raw materials it came from. Sugar cane, for example, is pressed into molasses, which is worth more than the cane; the molasses is processed into sugar and rum. The rum is refined and made up into several liquors and spirits, gaining value at each stage. A bottle of flavoured rum may cost the equivalent of up to US$20; its production required sugar, rum, flavouring and several other products, which in their unprocessed state may have cost less than $1.

The difference between the cost of the manufactured product and the raw materials from which it was derived is the 'value added'. This value is added as a result of the expensive outlay involved in building the factory and other expenses such as insurance, fuel taxes, etc.

Secondary economic activities in the Caribbean

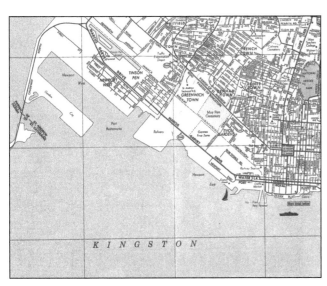

▲ *Figure 13.1 Port of Spain's coastline is dominated by industrial units*

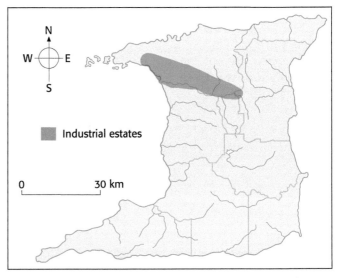

▲ *Figure 13.2 Industrial estates west and east of Port of Spain, Trinidad*

Industrial location is influenced by several factors that collectively contribute to the success of any industry. While they are all important, modern technology, government policies and international influence have changed the significance of any one factor.

- **Physical:** most industrial activities occur on relatively flat or undulating land. This is needed for the construction of the infrastructure and the easy movement of equipment and machinery during production. Some territories have used reclaimed land in coastal areas to establish industrial areas. Nearness to ports is important for easier import/export activities. Where industrial activity occurs inland there should be easy and relatively cheap movement of products to and from ports.

- **Human:** the availability of a labour force is important. Factories generally require both skilled and unskilled workers, although some factories may require more skilled and technologically apt labourers.

- **Economic:** industrial activities require a large capital input for both start-up capital (development of infrastructure) and operation costs. Raw materials, fuel and labour require the most expenditure for operation.

Garment industry in Trinidad and Tobago

Driven by the energy sector, Trinidad is the largest producer of manufactured products in the Caribbean. The 'industrialisation by invitation' model in the post-independence period was used to provide non-agricultural employment for its population and created the shift from a dominant primary economic activity economy to the now strong manufacturing-oriented economy. This resulted in many overseas garment manufacturers using the incentives provided on these estates to establish economic activities on the island. More recently, indigenous brands such as Emotions, Infinity, Generations and others have been produced for local and export markets.

Several factors have influenced the location of the garment industry in Trinidad and Tobago.

- **Physical:** the garment industries are located on industrial estates such as Diamond Vale in the west to 'Trincity' in the east – the so called 'East–West Corridor' between Port-of-Spain and Arima. The area comprises low-lying land between the Northern Range and Caroni Swamp east of the capital. Some companies are located outside these estates but generally they are in the area from Port-of-Spain eastwards to Arima, relatively close to the port area of Port-of-Spain and easily accessible by good roads. A wide variety of clothing is made in Trinidad: babywear, uniforms, lingerie and men's and women's garments. There is also a related fashion design industry including such designers as Meiling.

- **Human:** the industrial corridor lies between two of the large population centres – Port-of-Spain and Arima – and the west coast itself is more populated than any other geographical area. The Faculty of Engineering of the University of the West Indies is located in Trinidad. Highly skilled personnel from around the region are trained here and many are employed in the manufacturing sector after graduation. Trinidad has a skilled labour force that is largely computer literate and technologically aware. This is a product of accessible education through various institutions, which aims specifically to give young people technology training. Labour costs are also relatively low in Trinidad. A large number of unskilled workers are also employed.

- **Economic:** Trinidad has large reserves of natural gas, which is piped from supplies south of the industrial estates, so electricity is readily available and cheap. The government has provided incentives through tax exemptions on the import of raw materials such as cotton yarns. It has also contributed to the development of industrial estates, which are all equipped with electricity, water, telephone and waste disposal services. The world-class infrastructure has been influenced by a modern construction industry and foreign investors. In Trinidad, international investment by transnational corporations (TNCs), influenced by legislation (law) which allows foreigners to own and manage their own operations, has contributed significant capital to the establishment of garment and other industries.

✳ **Transnational corporations (TNCs)**, also called multinational companies (MNCs), are international companies that have global investments, worth billions of dollars, in several different countries. The parent company is usually from a more developed country (MDC). TNCs help in economic development because they have economic and political power and access to capital, labour and technology.

Trends ...

- **Increased production:** with the Fredrick Settlement Free Zone, Wallerfield industrial estate and the University of Trinidad and Tobago as an added source of skilled labour, more garment manufacturers will be attracted to the country.

- **Widening market:** Trinidad has expanded its markets to as far as Asia. Cheap energy allows Trinidad to deliver goods at competitive prices even compared with foreign imports. It should benefit from the CSME as it widens its markets.

... and challenges

Trinidad's garment industry is doing well at present, but there are still some challenges which, if they are not addressed, could have a significant effect on the industry. One such challenge is the import of cheap clothing that is usually brought in by informal traders. These retailers bring in cheap clothing in suitcases, usually from the USA. The garments are sold cheaply on the roadside. These retailers have few overhead expenses, so it is difficult for local manufacturers to compete with them.

➤ *Figure 13.3 Sidewalk sales in Trinidad*

Garment industry in Singapore (a newly industrialised country)
Singapore is the smallest country in Asia and one of the smallest in the world, but it stands out from others because it has a strong economy. Singapore has no natural resources but it has become rich in terms of GDP per head. Several factors have contributed to the growth and development of its garment industry.

● **Physical:** Singapore is strategically located on the Malacca Strait at one of the world's main crossroads of transportation routes, close to other industrialised and developed countries. It has large, technologically advanced port facilities. It has also been an entrepot port – that is, one that collects goods from other countries in the region and trades them on the world market. For example, Singapore purchases most of its raw materials from India and Indonesia. It trans-ships some of them elsewhere, but a large part is used in the local manufacture of designer clothing. Partly made clothing parts are also shipped to Singapore for assembly.

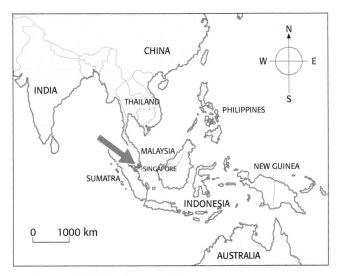

▲ *Figure 13.4 South East Asia*

▲ *Figure 13.5 Port facilities at Singapore*

- **Human:** the country has an educated and highly skilled workforce, most of which is engaged in manufacturing. Labour is also relatively cheap.

- **Economic:** Singapore is wealthy, so it can inject capital into the manufacturing sector. It buys raw materials, like oil, and processes these into various products – iron and steel and car parts, for example – and then processes and assembles these into finished products. This has contributed to the growth of its auto industry. Agricultural products are processed and shipped all over the world.

Singapore was classified as a newly industrialised country (NIC) and one of the 'Asian Tigers' because of its increasing share of the world's manufacturing output. It has recently been redefined as a more economically developed country (MEDC). This is partly in terms of the growth in other sectors, such as the tertiary and quaternary sectors.

➤ *Table 13.1 Singapore –*
employment structure by
industry

	2002 (thousands)	2003 (thousands)
Total labour force (active)	2 017.4	2 033.7
Manufacturing	367.6	364.8
Construction	119.1	114.5
Wholesale and retail	304.4	296.4
Hotel and restaurant	125.3	128.4
Transport storage and communications	218.8	216.0
Financial services	107.9	104.7
Business services	237.4	243.0

Source: ILO Yearbook of Statistics

➤ *Table 13.2 Trinidad –*
employment structure by
industry

	2002 (thousands)
Total labour force (active)	586.2
Sugar	11.2
Agriculture, fishing, forestry, hunting	24.9
Petroleum and gas (including production, refining and services)	17.2
Other mining and quarrying	0.8
Other manufacturing	55.8
Construction	68.9
Wholesale and retailing, restaurant and hotel	94.6
Transport, storage and communications	41.9
Finance and business	43.8
Community (government), social and personal	158.1
Not stated	1.4

Source: ILO Yearbook of Statistics – Caribbean labour statistics

Go to www.ILO.org for more information on the employment structure in other Caribbean territories and other MEDCs.

Exercise

Use Tables 13.1 and 13.2 to answer the following questions.

1 In which sector/s were the largest numbers of the labour force employed in Singapore in 2002?

2 In which sector/s were the largest numbers of the labour force employed in Trinidad in 2002?

3 Using the figures for 'Total active labour force', calculate the percentage for each sector for 2002. Create a divided circle (pie chart) to show your results.

4 What sectors are similar in structure between Trinidad and Singapore?

5 What economic activity is not accounted for in Table 13.1? Give reasons for your answer.

6 What sector experienced the largest percentage increase in employment in Singapore between 2002 and 2003, and by what amount?

7 List the sectors that experienced a decline in employment between 2002 and 2003 in Singapore.

8 Describe differences in the structure of employment in Trinidad and the structure of employment in Singapore in 2002. Refer to the figures in your answer.

Food processing in Trinidad and Tobago

Food processing or food manufacturing is directly linked to the agriculture industry. Fresh farm produce cannot last very long without some form of preservation or processing. The processing of foods adds shelf life to fresh produce, and increases its potential market over a longer time. The food processing industry has grown throughout the Caribbean as research and technology have developed new ways of diversifying our foods, from preserving spices to making exotic liquors. Trinidad is now the top producer of processed foods in the Caribbean.

➤ *Table 13.3 Some processed foods from the Caribbean*

Country	Products (processed food)	Raw materials (local)
Barbados	Rum, yoghurt	Sugar cane, milk
Jamaica	Rum, jerk sauce	Pepper and spices
Trinidad and Tobago	Canned juice, ketchup	Citrus fruits, tomatoes

Exercise

Create your own table and see how many other Caribbean countries and products you can add to this list.

The original food processing activities in Trinidad and Tobago involved the processing of coconut, citrus fruits, sugar, rum and bitters. These were made primarily from materials produced locally. Citrus and other fruits, for example, are processed into concentrates, preserves, juices and essential oils. However, as Trinidad's manufacturing sector grew, newer industrial estates introduced a wider variety of processed foods, many of which are made from imported raw materials. Some of these include confectionery, condiments, canned and packaged vegetables, rice and macaroni products and flour. The country also produces large quantities of frozen and processed fish, chicken, beef and pork products.

➤ *Figure 13.6 Food processing plants in Trinidad*

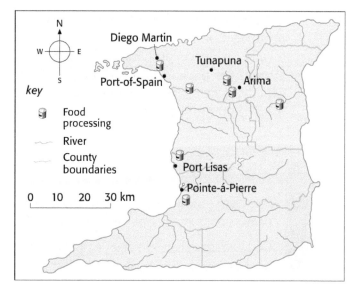

Several factors have influenced the location of food processing industries.

- **Physical:** like other manufacturing industries in Trinidad, the food processing facilities are located on flat land, some near to port facilities and others near to the source of raw materials. Sugar processing and rum distilling plants are located on sugar estates where the cane is readily accessible. Most other food processing is along the industrial corridor between Port-of-Spain and Arima, and along the coast in the town of San Fernando.

- **Human:** labour is readily available, both skilled and unskilled.

- **Economic:** Trinidad's relatively cheap energy supply has boosted the growth of industrial estates and this has attracted both regional and international investors. Government incentives are also very attractive.

Trinidad's food processing sector is growing rapidly and is by far the largest in the Caribbean. The industry now imports a large proportion of its raw materials from other Caribbean countries as well as North America, Europe, Asia and Africa. Catelli Primo, for example, produces pasta products using flour from Canada, pigeon peas imported from Nigeria, ketchup with tomato paste from Spain, and orange juice with concentrate from Belize. Some large companies from within the Caribbean region have moved their headquarters and operations to Trinidad. Trinidad is now producing food for the entire Caribbean region.

However, the rapid growth in the food processing industry has created a greater demand for raw materials than farmers can supply, so imported products have to be used. Variations in the quality of raw materials from different suppliers can affect the quality of the finished product. Competition from imported products also poses a threat to locally produced processed foods. The imports are sometimes more attractively packaged and more intensively marketed through expensive media advertising. They therefore appeal to consumers more than the local products.

Exercise

This involves collecting research data. You will first need to acquire permission from your parents and a letter from your class teacher.

Go to your local supermarket (your largest store) and look for the local food products (canned or bottled). Conduct a survey using the labels on the products to find out how much of it was 'Made in Trinidad and Tobago'.

Tertiary economic activities – tourism

Tourism is not only the dominant tertiary activity in the Caribbean region but it is also now the major economic activity in most Caribbean countries. In fact, the region as a whole is one of the most tourism-dependent regions in the world. The primary elements for tourism development in the Caribbean are simple: sand, sea and sun. The warm climate, clear blue waters and predominantly white sandy beaches have contributed to the 'three Ss' concept. However, the region also markets other elements as 'adventure tourism', such as its interior natural landscape and rich cultural heritage. This trend is centred on a sustainable approach to tourism.

The location of tourism in the Caribbean

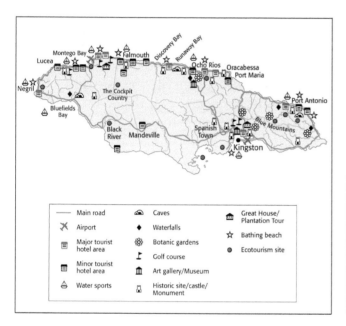

▲ *Figure 13.7 Tourism activities in Jamaica*

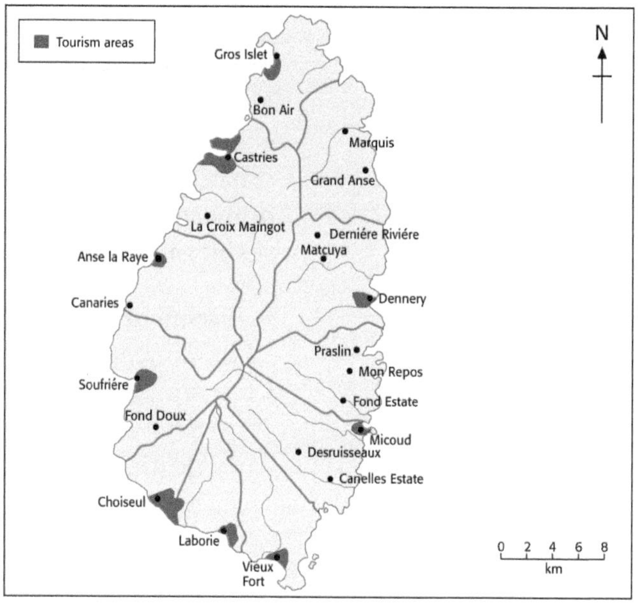

▲ *Figure 13.8 Tourism activities in St Lucia*

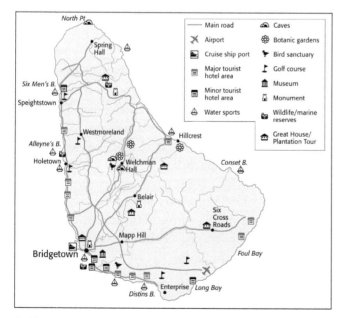

▲ *Figure 13.9 Tourism activities in Barbados*

▲ *Figure 13.10 Tourism activities in The Bahamas*

Tourism in The Bahamas

The factors contributing to tourism development in the region are common to many places, primarily because the climate and geography are similar throughout the Caribbean. The variations and diversity come about through the uniqueness of each territory's culture and its approach to tourism development and management.

The Bahamas consist of a group of nearly 700 islands and cays (small islands) and more than 2 000 low, barren, coral banks. Only about 22 of the islands are inhabited – New Providence and Grand Bahamas are home to 75 per cent of the Bahamian population. The Bahamian economy is centred on tertiary activities, primarily tourism and offshore banking. Tourism accounts for nearly half the country's gross domestic product (GDP). The success of tourism in the Bahamas is a result of several factors.

● **Physical features:** The Bahamas is an ideal example of the 'three Ss' concept:
  ● sun: the warm subtropical marine climate with mildly cool but dry winter months makes The Bahamas an ideal tourist destination, especially for people from the north who want to get away from the bitter winter cold (the average annual temperature is 25°C)
  ● sand: the coral banks have created an environment for the development of large stretches of white sand beaches with gentle breaking waves
  ● sea: crystal-clear waters.

● **Geographic location:** The Bahamas is also ideally located to attract the largest tourist market, the United States of America. It is the closest Caribbean territory to the USA, being only 40 minutes by air from Miami international airport. The islands are linked by air to many US cities, especially those along the east coast. There are about 50 airstrips that can accommodate various sizes of aircraft. Cruise ships make daily calls to Freeport and Nassau, and the area is also popular as a yachter's paradise.

● **Human factors:** most people in The Bahamas are involved in tourism or tourism-related economic activities. Many migrants from other Caribbean territories are employed in the sector. Most jobs are in the many hotels and restaurants.

● **Economic:** the government of The Bahamas has given special tax concessions to developers, especially of large tourist resorts. The Bahamas has some of the largest luxury resorts in the region. The infrastructure is well-established and extensive. Advertisement and management of tourism is carried out by the tourist board. The

▲ *Figure 13.11 Position of The Bahamas in relation to the USA*

Bahamas accepts nearly 2 million visitors annually and so a multiplier effect is in action. Offshore banking and financial activities also help to boost the tourism industry.

▲ *Figure 13.12 Atlantis in The Bahamas*

Country	Tourist arrivals (2004)
Bahamas	1450043
Barbados	551953
Dominica	30988
Dominican Republic	3443205
Grenada	89854
Jamaica	298431
St Lucia	132039
Turks and Caicos	1414786

▲ *Table 13.4 Stopover tourists arrive in the Caribbean*

Trends in tourism in the Caribbean

In an effort to enhance tourism and present the Caribbean as the number one destination in the world, Caribbean countries are moving towards 'sustainable tourism'. It is now understood that this form of tourism also protects the 'raw material' – the environment that attracts the tourist in the first place. Sustainable tourism includes a variety of attractions, for example eco, adventure and heritage sites.

✳ **Sustainable tourism** has as its main focus the proper management and development of natural, cultural and human resources on a basis that sustains them for future generations. The concept aims at developing unique experiences for visitors and improving the quality of life of all citizens and communities involved in tourism activities.

✳ **Ecotourism** is a specialised form of tourism that uses untouched natural environments (plants, animals, soil and general topography) as an attraction for visitors. The idea is to leave untouched the ecology of these areas, which are designated as protected areas. The natural state becomes a fascination for tourists who also want to help to protect the environment.

✳ **Heritage tourism** is based upon historical and cultural themes that involve the visiting of sites such as museums, special buildings and traditional rural communities. Heritage tourism also involves major events such as festivals and celebrations that are based on cultural themes.

Since the 1990s, ecotourism has been recognised as a viable form of sustainable tourism development (Table 13.5).

➤ *Table 13.5 Some examples of ecotourism and heritage sites and events in the Caribbean*

Belize	<ul><li>Central America's paradise of reefs – second largest barrier reef in the world</li><li>Vast tracts of undisturbed rainforests in the Cayo District</li><li>Mayan sites – Caracol, one of the largest in Central America</li></ul>
Barbados	<ul><li>Majestic underground cave systems, e.g. Harrison's Cave</li><li>Old archeological sites along the north and north-east coasts</li><li>Events such as Crop Over and Oistin's Fish Festival – significant heritage events</li></ul>
Dominica	<ul><li>Thick rainforest</li><li>Carib reserve</li><li>Sulphur springs in Valley of Desolation</li></ul>
Jamaica	<ul><li>Spectacular Cockpit Country</li><li>Black River Morass</li><li>Blue and John Crow Mountain National Park (BJMNP), the only National Park in Jamaica. It includes the famous Blue Mountain Peak Hike, Cunu-Cunu Pass (home of the second largest butterfly in the world) and the Holywell Recreational Grounds equipped with a Kids' Discovery Zone interpretation and play centre</li><li>Heritage of the Maroons in the Rio Grande Valley</li><li>Birthplace of reggae icon Robert Nesta (Bob) Marley</li></ul>
St Kitts	<ul><li>Largest crater in the Caribbean: 853 m volcanic rim, nearly 2 km wide</li><li>Brimstone Hill Fort World Heritage Site</li></ul>
St Lucia	<ul><li>Soufrière heritage sites including the Pitons (Gros Piton and Petit Piton) now a World Heritage Site, the caldera with bubbling sulphur springs, and the rainforests</li><li>Diamond Falls</li></ul>
Trinidad and Tobago	<ul><li>Buccoo Reef off the coast of Tobago has the largest brain coral in the world</li><li>Tobago has the oldest forest reserve in the western hemisphere</li><li>Pitch Lake in La Brea, Trinidad</li><li>Trinidad is known worldwide for its steelband music and carnival</li><li>Birds of Caroni Swamp</li></ul>

➤ *Figure 13.13 A natural tourist attraction of the Caribbean*

The Caribbean has become one of the most tourism-dependent regions in the world. While this is great news for economic development there are some challenges too. At its Seventh Annual Sustainable Tourism Conference in Tobago in 2005, the Caribbean Tourism Organisation (CTO) cited some of them.

● 'Maintaining the tourist flows necessary to guarantee economic stability': while an increase in the number of visitors increases income, too many tourists can deplete resources that in some cases are already in limited supply for locals. Also, larger numbers do not always mean more income. Large cruise ships dock at Caribbean ports but many of the tourists either stay on board or, if they do not disembark, spend very little in the local shops.

● 'Ensuring proper use of resources for the benefit of both locals and visitors': in many instances tourism development has overlooked the needs of local people. For example, there may be large golf courses with water fountains and sprinklers in or near communities while the local people have little or no access to domestic water.

● 'Ensuring that the resources which now attract visitors to the Caribbean will continue to exist and attract them in years to come': the relationship between tourism development and environmental conservation is important and must be considered in all developments.

There are other challenges, too.

- **Competition from other global destinations:** the Caribbean islands must compete with larger and cheaper destinations in, for example, the Americas and the Mediterranean. Some of the developments in these regions are owned by large multinational corporations or rich tycoons who can afford to buy cheap land and develop 'a human paradise'. There are also other tropical destinations that offer similar packages to the Caribbean, such as Fiji and Tahiti.

- **Social issues:** poverty, crime and violence are still problems in some Caribbean territories; visitors do not want to have to deal with such domestic issues while they are on holiday. Even if crime and violence do not affect tourists, especially when they are enclosed in an all-inclusive resort, they may want the option to explore the country without fear of getting hurt or having their idea of paradise spoilt by the sight of poverty (shanties, street people, run-down streets).

- **Seasonal employment:** this has always been an issue for those involved in tourism-related economic activities. Some larger hotels lay off their staff or reduce their working hours during the 'off-season'.

- **Globalisation:** this affects tourism development because it challenges the culture and heritage of individual countries. It is difficult to market local products when foreigners import their own, and impose their own taste on the economy. The demand for certain products means that potential foreign exchange earned is spent on imported items.

- **Export of profits:** much of the money earned in tourism is created and remains in foreign markets. For example, the bookings and payments are made overseas, so the foreign exchange stays in those countries. The money made from large international hotels and also some locally owned ones is invested in foreign accounts, so potential earnings go back into foreign currency reserves.

- **Transport:** the Caribbean currently has two international carriers – Air Jamaica and BWIA – both of which have financial problems. Connecting flights within the region can be limited and some territories still have limitations on the size of aircraft that can be accommodated there.

➤ *Figure 13.14 Passengers delayed at Piarco International Airport, Trinidad and Tabago*

For tourism to be successful, it must maintain a balance in its economic, environmental and social development. The Caribbean countries must decide, therefore, whether they want tourism *growth* or tourism *development* (see Chapter 15).

Exercise

Conduct a class debate on sustainable tourism.

Split into groups to debate the following issues:
● The Caribbean region needs tourism growth not tourism development.
● Ecotourism is a viable form of tourism suitable for economic development.
● The Caribbean region should have one single 'brand' of tourism, rather than several brands.

Go to www.onecaribbean.org the Caribbean Tourism Organisation (CTO) for information on regional tourism policies and development.

Exam practice questions

Human systems

Paper 1: Multiple choice

1 Which of the following statements are reasons why large numbers of tourists from North America visit the Caribbean?
(i) North America is close to the Caribbean.
(ii) Air and sea transportation to the Caribbean is abundant.
(iii) North America is in the New World.
(iv) The Caribbean is hot all year round.
A All of the above
B (i), (ii) and (iii)
C (i), (ii) and (iv)
D (i), (iii) and (iv)

2 The Caribbean needs tourism mainly to:
A Replace sugarcane cultivation in the region
B Provide a use for beautiful beaches
C Enable citizens to see how foreigners behave
D Provide foreign currency for the region

3 Which of the following statements are true of sustainable tourism?
(i) Communities are involved.
(ii) Hotels are built on beaches.
(iii) Management of resources is encouraged.
(iv) Ecotourism is encouraged.
A (i), (ii) and (iv)
B (ii), (iii) and (iv)
C (iii) and (iv)
D (i), (ii) and (iv)

4 Which of the following are non-renewable resources?
A Oil and forest
B Forests and streams
C Bauxite and oil
D Bauxite and fisheries

5   Which of the following could be best described as secondary economic activity?
   A Tourism
   B Food processing
   C Bauxite mining
   D Banking

6   The governments of most countries in the Caribbean would like to intensify the region's agriculture. This means they want to:
   A Use more land for farming
   B Engage most of the labour force in agriculture to get more yield per hectare
   C Plant more traditional crops
   D Diversify crops and get more yield per hectare of land

7   Which of the following considerations is *most* important when choosing a location for mining?
   A A large labour force
   B Nearness to raw material
   C Adequate marketing facilities
   D Adequate storage facilities

8   Which of the following is the main reason why many factories in the Caribbean are situated in coastal locations?
   A Availability of cheap labour
   B Availability of cheap power supply
   C Nearness to waste disposal facilities
   D Nearness to facilities for unloading raw materials

9   Which of the following are the main reasons for urbanisation in the Caribbean?
   (i)   Natural increase of the population in the cities
   (ii) Expectation of more jobs in the cities
   (iii) Migration from urban to rural areas
   (iv) Migration from rural to urban areas
   A (i) and (iii)
   B (i) and (iv)
   C (i), (ii) and (iii)
   D (i), (iii) and (iv)

10 Many garment factories in the Caribbean are closing mainly because:
   A More people travel to North America to shop
   B Cheaper imports are entering the region
   C The rates for wages are higher in the Caribbean than in other regions
   D Many women refuse to work in the factories

11 Which of the following statements best describes the rapid growth in the size of most urban centres in the Caribbean?
A An ageing population
B The migration of people from rural areas
C High birth rates and low death rates
D An increase in the number of immigrants

12 Which of the following areas would have the lowest population density in the night?
A A squatter settlement
B An apartment complex
C A suburban residential area
D A central business district

13 What factor has not contributed to the location of the manufacturing processing industry in the Caribbean?
A Availability of cheap labour
B Availability of raw material
C Cheap imports
D Nearness to markets

➤ *Figure P.1*

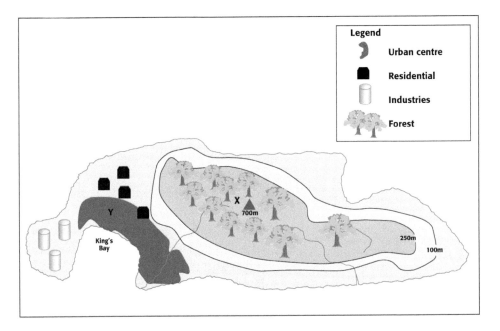

14 Which of the following features on the sketch map above best explains why the urban centre is located at Y?
(i) The presence of flat land
(ii) The nearness to the mountains
(iii) The presence of a natural harbour
A (i) and (ii)
B (i) and (iii)
C (ii) and (iii)
D (i), (ii) and (iii)

Total 14 marks

Paper 2

1 (a) Study the table below showing the origin of visitors to the Caribbean during a particular year.

➤ *Table P.1 Approximate tourists arrival in the Caribbean (thousands)*

	from North America	**from Europe**
Bahamas	1500	175
Barbados	200	650
Jamaica	900	200
St Lucia	550	125

    (i) How many visitors went to Jamaica from North America that year?

    (ii) Name one of the main destinations of visitors from North America.

    (iii) Name the main source of visitors to Barbados.

    (iv) What is the approximate total of visitor arrivals to Jamaica?

                     (4 marks)

(b) (i) List THREE ways in which all-inclusive resorts can do more harm than good. (3 marks)

    (ii) Describe TWO ways in which tourism affects the environment. (4 marks)

(c) (i) Explain the difference between primary and secondary industry. (4 marks)

    (ii) Explain how each of the following factors influences the growth of tourism industry in the Caribbean:

        (a) transportation

        (b) social and political conditions. (6 marks)

Total 24 marks

    (iii) Define the term 'ecotourism'. (3 marks)

➤ *Figure P.2*

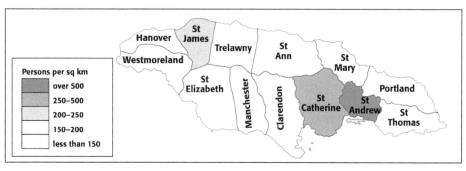

2(a) Study the choropleth map in Figure P.2, which shows population distribution in Jamaica, and answer the following.

    (i)   What is the population in Montego Bay?

    (ii) What is the highest population density?

    (iii) How many times is the population of St Catherine larger than that of Trelawny?   **(3 marks)**

(b) Describe THREE factors that contribute to population distribution in a named Caribbean country.   **(6 marks)**

(c) (i)   What is meant by the term 'urbanisation'?   **(3 marks)**

    (ii) Describe TWO factors that influence the selection of the site of the capital city of a named Caribbean country.   **(4 marks)**

(d) Compare the growth and development of *either* New York City or Tokyo with that of a named Caribbean capital, under the following headings:

    (i)   Location

Total 24 marks   (ii) Functions   **(8 marks)**

3   Study the table below and then answer the following questions.

➤ *Table P.2 World energy consumption, 1998*

Activity	Percentage of energy consumed
Industry	40
Transport	28
Residential/commercial	32

(a) Construct a divided circle to represent the information given in the table above.   **(4 marks)**

(b) Name ONE type of primary economic activity and one type of secondary economic activity in the Caribbean.   **(2 marks)**

(c) Within recent years food processing has increased in some Caribbean territories.

    (i)   Give TWO reasons why this economic activity has been located in a **named** Caribbean country.   **(6 marks)**

    (ii) Describe THREE problems faced by the food processing industry in the Caribbean.   **(6 marks)**

(d) Describe THREE trends in secondary economic activity in
Total 24 marks   Singapore with trends in a named Caribbean territory.   **(6 marks)**

CHAPTER 14

Natural hazards

By the end of the chapter students should be able to:

- define the term **natural hazard**
- describe the impact of one of the following on life and property: **volcanic eruptions**, **earthquakes** or **hurricanes**
- explain the **response to natural hazards** in a named Caribbean country at individual, national and regional levels.

▲ *Figure 14.1 Destruction caused by Hurricane Ivan*

Natural hazards are extreme natural events that have the potential to threaten both human life and property. Extreme events in unpopulated areas are not hazards. In this chapter we look at geophysical hazards: **volcanic eruptions**, **earthquakes** and **hurricanes**. These three hazards result in great loss of life and damage to property in the Caribbean and around the world.

❋ A **natural hazard** is an extreme event that threatens human life and property.

Impact of natural hazards

The impact of a hazard on a country depends on the:

- severity of the event
- density of population
- knowledge and preparedness of the population
- prediction/forecasting technology available.

Larger, wealthier countries often have more resources dedicated to lessening the impact of hazards. For example, in the USA each type of hazard has its own organisation and regional group to monitor it, while in the Caribbean there are fewer hazard organisations.

The impact in less developed countries, like those of the Caribbean, is often higher in terms of loss of life but lower in terms of the value of property damage than in more developed countries like the USA. However, the fragile economies of the Caribbean take a much longer time to recover than devastated areas of the USA, which qualify for Federal Assistance.

See www.disastercenter.com for more information on hazard events.

Impact of volcanic eruptions

Volcanoes which erupt with lava flowing quietly down the sides of the volcano are not usually hazardous since the impact of their eruption can be avoided, as for example in Hawaii. Volcanoes such as Mauna Loa erupt continuously but have little impact on human activity.

In the Caribbean, however, volcanoes erupt explosively. These eruptions have several hazardous features:

- lava flows

- toxic gases

- pyroclastic flows

- ash falls.

Each of these can result in loss of human life and property, mainly through burning or suffocation (see Chapter 4).

The impact of volcanoes on Caribbean life and property has been devastating throughout history. Even on non-volcanic islands like Barbados, ash falls from neighbouring islands can damage roofs.

The eruption of Mt Pelée on Martinique in 1902 destroyed the town of St Pierre. A tsunami killed 100 people and a pyroclastic flow of hot gases travelling very fast killed 30 000 people in just two minutes. The town was completely destroyed.

The 1997 eruption of Langs Soufrière on Montserrat resulted in the total destruction of the south of the island. The capital, Plymouth, was burnt to the ground and submerged in ash and lava. All the buildings and vegetation were destroyed. The main airport and seaport were also rendered useless. Large lava and pyroclastic flows masked the landscape and killed nine people who had remained in the evacuated area. The entire infrastructure of the island and all economic activities had to be recreated in the north of the island.

▲ *Figure 14.2a Plymouth, Montserrat, before the eruption*

▲ *Figure 14.2b Plymouth, Montserrat, after the eruption*

Impact of earthquakes

Earthquakes create many hazards. They are capable of totally wiping out human life and property.

- **Tremors** produced by earthquakes threaten all built structures and natural features standing above the Earth's surface. These may collapse, burying the people inside them.

- **Ground fissures** can break pipelines, roads and bridges and cause fires in the process.

- **Tsunamis** caused by underwater earthquakes can sweep away coastal settlements.

▲ *Figure 14.3 The Indian Ocean earthquake and tsunami, 26 December 2004*

The Asian tsunami of 26 December 2004 was referred to by the Secretary General of the United Nations as an 'unprecedented global catastrophe'. The giant, fast-moving wave affected countries from Sumatra just east of the epicentre, through the coast of India and the Maldives to Somalia in East Africa. It washed away entire villages along the coast of northern Sumatra including Aceh, and nearby areas of southern Thailand. It caused 290 000 deaths and millions of people were made homeless. Damage in Aceh alone was estimated at US$2 billion, as the agriculture and fishing sectors were devastated.

Earthquakes are common in the Caribbean, especially along the active convergent and transform plate margins (see Chapter 4). Historically, the Caribbean has experienced many smaller tsunamis associated with volcanic eruptions or submarine earthquakes. The most serious potential hazard is the tsunami that would result from a major eruption of the underwater Kick 'em Jenny volcano near Grenada.

➤ *Figure 14.4 Caribbean Earthquakes*

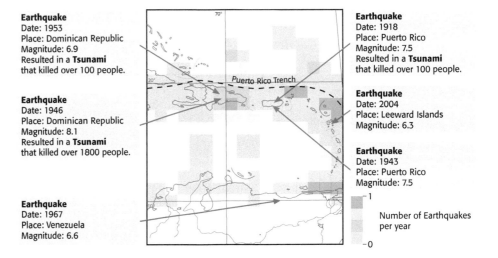

Earthquake
Date: 1953
Place: Dominican Republic
Magnitude: 6.9
Resulted in a **Tsunami** that killed over 100 people.

Earthquake
Date: 1946
Place: Dominican Republic
Magnitude: 8.1
Resulted in a **Tsunami** that killed over 1800 people.

Earthquake
Date: 1967
Place: Venezuela
Magnitude: 6.6

Earthquake
Date: 1918
Place: Puerto Rico
Magnitude: 7.5
Resulted in a **Tsunami** that killed over 100 people.

Earthquake
Date: 2004
Place: Leeward Islands
Magnitude: 6.3

Earthquake
Date: 1943
Place: Puerto Rico
Magnitude: 7.5

Number of Earthquakes per year

Exercise

Study Figure 14.4.

1 In which year was the strongest earthquake felt in the Caribbean?

2 Which areas experience earthquakes most often?

3 Which country has suffered the greatest earthquake damage?

4 Describe the distribution of tsunamis. How does this relate to the plate margins?

The Mercalli modified intensity scale (Table 14.1) classifies earthquakes according to the effects of ground vibration and structural damage.

▼ *Table 14.1 Mercalli modified intensify scale*

Intensity value	Description
I	Not felt except by a very few under especially favourable circumstances.
II	Felt only by a few persons at rest, especially on upper floors of buildings. Delicately suspended objects may swing.
III	Felt quite noticeably indoors, especially on upper floors of buildings, but many people do not recognise it as an earthquake. Standing automobiles may rock slightly. Vibration like passing of truck.
IV	During the day felt indoors by many, outdoors by few. At night some awakened. Dishes, windows, doors disturbed; walls make creaking sound. Sensation like heavy truck striking building. Standing automobiles rock noticeably.
V	Felt by nearly everyone, many awakened. Some dishes, windows and so on break; cracked plaster in a few places; unstable objects overturned. Disturbance of tall trees, poles and other objects sometimes noticed. Pendulum clocks may stop.
VI	Felt by all, many frightened and run outdoors. Some heavy furniture moved; a few instances of fallen plaster and damaged chimneys. Damage slight.
VII	Everybody runs outdoors. Damage negligible in buildings of good design and construction; slight to moderate in well-built ordinary structures; considerable in poorly built or badly designed structures; some chimneys broken. Noticed by people driving cars.
VIII	Damage slight in specially designed structures; considerable in ordinary substantial buildings with partial collapse; great in poorly built structures. Panel walls thrown out of frame structures. Fall of chimneys, factory stacks, columns, monuments, walls. Heavy furniture overturned. Sand and mud ejected in small amounts. Changes in well water. People driving cars disturbed.
IX	Damage considerable in specially designed structures; well-designed frame structures thrown out of plumb; great in substantial buildings, with partial collapse. Buildings shifted off foundations. Ground cracks conspicuously. Underground pipes break.
X	Some well-built wooden structures destroyed; most masonry and frame structures destroyed, along with foundations; ground badly cracked. Rails bent. Landslides considerable from river banks and steep slopes, sand and mud shifts, water splashes, slopping over river and lake banks.
XI	Few, if any, (masonry) structures remain standing. Bridges destroyed. Broad fissures in ground. Underground pipelines completely out of service. Earth slumps and landslips in soft ground. Rails bent greatly.
XII	Damage total. Waves seen on ground surface. Lines of sight and level distorted. Objects thrown into the air.

Impact of hurricanes

- High velocity **winds over 120 km/h** (75 mph) are the major threat – they can take off roofs and even carry people into the air.

- **Storm surges** may flood coastal areas.

- **Torrential rain** accompanying hurricanes can result in flooded valleys.

- **Landslides** may push down and bury houses and block roads.

In 2004 Hurricane Ivan did terrible damage in many islands of the Caribbean and parts of the coast of the USA.

The 2005 season was very active with more than 20 named storms. Hurricanes Katrina and Rita battered New Orleans and the southern USA.

Impact of Hurricane Ivan in Grenada

Hurricane Ivan struck Grenada on 8 September 2004. It was a Category 3 hurricane (see Table 14.2) with maximum sustained winds of 195 km/h (120 mph). The eye passed directly over the island, and the hurricane caused substantial damage:

- 22 reported deaths (drowned sailors, and people crushed in collapsed homes and under trees)

- estimated US$1 billion in damage

- 8 000 injured and traumatised homeless people given shelter in the few undamaged schools

- nine out of ten buildings lost their entire roofs and/or had collapsed or damaged walls, including some shelters and public buildings such as the prison, schools, churches, the Prime Minister's residence, the Emergency Operation Centre building, hotels

- disruption of water, telephone (including cell phones) and electricity supplies, radio/TV communications

- looting and security problems (e.g. all prisoners set loose)

- complete destruction of agriculture and tourism sectors – loss of jobs

- closure of airport and seaport

- roads blocked with fallen trees, poles and power lines

- little food available following the hurricane.

▲ *Figure 14.5 Hurricane Ivan roars through the Caribbean*

Exercise

Look at each photograph in Figure 14.5 and identify and describe at least two different types of damage you can see.

Hurricanes are categorised by wind speeds and expected levels of damage. The Saffir–Simpson scale is used to describe Atlantic hurricanes (Table 14.2).

➤ *Table 14.2 Simplified Saffir–Simpson scale*

Category	Max. winds km/h (mph)	Damage
1 Minimal	118–152 (74–95)	Vegetation, mobile homes affected
2 Moderate	154–176 (96–110)	Doors, windows affected, coastal flooding
3 Extensive	178–208 (111–130)	Collapsed walls, floating debris
4 Extreme	210–248 (131–155)	Roof/building destruction, beach erosion
5 Catastrophic	>248 (>155)	Complete building failure – blown away

Exercise

1 Use the following headings to create a table of damage done to life and property by each of type of hazard: volcano, earthquake, hurricane.

- Injury/death
- Water supply
- Food supply
- Homelessness
- Communication
- Infrastructure
- Economic/jobs

Set out your table like this:

Hazard	Impacts

You could work in groups, each group on a different hazard.

2 Imagine you are a leader of a small island for which a hazard warning has been issued. Explain the steps you would take to ensure the safety of your people. Issue a press release with this information.

3 Describe the response that the city of New Orleans took to Hurricane Katrina and the effect of the hurricane on the city.

Human response to natural hazards

✳ **Response** is the action taken by individuals, countries and regions before and after hazardous events.

Individual level

The obvious, sensible response is not to live in a hazardous area. However, the actual response lies in complex individual/group decisions.

- Hurricane Camille (1969) threatened the Mississippi region and evacuation orders were given. Twenty people stayed in the area for a 'hurricane party' and lost their lives. Was that decision reckless or daring?

- Bangladesh is a country in the low-lying fertile Ganges delta area where an average of 580 people live in each 5 km². Every year the country is devastated by hurricanes. In 1970, 300 000 people were killed, and 139 000 were killed in 1991. Yet there has not been any major relocation of people to safer areas. Why do they stay?

One major factor influencing the decision to live in hazardous areas is the beneficial aspects of these areas (Table 14.3).

➤ *Table 14.3 Positive aspects of areas with hazardous events*

Hazard	Advantages
Volcanoes	Fertile soils, precious minerals, new land (islands)
Earthquakes	Sunny, warm climate attracts many people for tourism
Hurricanes	Hot, moist climate suitable for agriculture, tourism

The response of individuals and groups to hazards varies according to a variety of factors, for example:

- What is their attitude to risk – apathetic acceptance of 'acts of God', or overconfident ('it can't happen to me')?

- Are alternative settlement areas available? Is evacuation to safer areas possible?

- Is the technology available to monitor and predict the hazard?

- Is there money available to prepare for the event and to lessen its impact?

- Do people have the knowledge and experience of the hazard in order to prepare for it?

There are two approaches to reducing the impact of hazards.

1 **Awareness of the hazard**: this includes volcano monitoring, earthquake prediction, hurricane forecasting.

2 **Control of the human activity threatened by it**: this includes land use zoning, construction regulations, education and preparedness.

➤ *Figure 14.6 Safeguarding a house against hurricane damage*

Individual response in Grenada

In Grenada in 2004, individuals were caught somewhat unprepared because the country had not been hit by a hurricane in recent years. Granada is a relatively poor developing nation, and many buildings were poorly built and susceptible to hurricane damage. One report stated that in the aftermath of the hurricane, people were walking about like zombies, stunned and dazed. Others used the chaotic situation to loot shops and houses.

National level

Most Caribbean countries have their own disaster response organisation. In Jamaica, it is the Office of Disaster Preparedness and Emergency Management of Jamaica (ODPEM), in Trinidad and Tobago it is the National Emergency Management Agency (NEMA), while in Barbados it is the Central Emergency Relief Organisation (CERO). Each of these organisations is responsible for preparing its own people to respond to their own hazard threat.

Go to www.disaster-info.net/carib/links to learn more about the organisation in your country.

National response in Grenada

Following the 2004 hurricane the government of Grenada quickly declared a state of emergency and immediately imposed a dusk-to-dawn curfew. It also sought help from the Caribbean Disaster Emergency Agency (CDERA) and the Regional Security System.

The national Emergency Operations Centre was itself handicapped by damage to its temporary buildings.

Regional level

The Caribbean Disaster and Emergency Agency (CDERA) was founded in 1991 as an intergovernmental organisation responsible for disaster management in 16 participating territories.

Go to www.cdera.org for more information on CDERA.

The University of the West Indies (UWI) also helps to keep member governments informed by researching and monitoring activity. For example, the UWI Seismic Research Unit based in Trinidad and Tobago maintains a 'volcanic surveillance and warning system' for volcano and earthquake-related activity in Trinidad and Tobago and the Eastern Caribbean.

Go to www.uwiseismic.com to find out about earthquake and volcano preparedness on each island.

The Caribbean Disaster Information Centre based at UWI, Mona provides a wide variety of information on disasters.

Regional response to Grenada

The regional response to the Grenada disaster operated at many levels.

- Immediate relief shipments to the island were organised, for example from Trinidad and Tobago, the sub-regional focal point of CDERA.

- Individual fishermen in Barbados and Tobago ferried water and cooked food to their neighbours in coastal Grenada.

- The Regional Security System sent detachments of soldiers and police to restore order.

- The Barbados Caribbean Broadcasting Corporation organised the only communication available for the Prime Minister of Grenada to address his country.

- Schools in nearby territories offered to accept Grenadian students so that they could continue their studies.

In the following weeks, crews of engineers and technicians helped to restore the infrastructure. Other private groups were involved, for example the students of the Barbados Samuel Jackman Prescod Polytechnic helped to repair the roof of Grenada College.

In an address to the Heads of Government meeting in Trinidad, Prime Minister Mitchell recorded his praise and gratitude for the region's help. He said he believed that the regional response to the Grenada disaster showed that regional unity was in fact a 'lived' experience.

Exercise

Work through the four steps of this project. The work will help to make your class and family safer.

1 Find out what hazard threats occur in your area.

2 Create a family disaster plan (make sure you involve all members of your family).

3 Keep an up-to-date checklist of what each person is to do when there is a hazard warning.

4 Practise your plan and modify it.

The completed plans could be prepared as class presentations.

CHAPTER 15

Pollution and global warming

By the end of the chapter students should be able to:

- define **pollution**
- describe **types of pollution**
- identify areas in the Caribbean where **pollution is a problem**
- descibe **measures used to reduce the harmful effects of pollution**
- define **global warming**
- describe the **long-term changes** in global temperature
- describe and explain **causes** and **consequences** of global warming in a named Caribbean country and in either Mauritius or the Maldives
- explain **measures** taken to reduce the impact of global warming in a more developed country.

▲ *Figure 15.1 Toxic chemicals and fumes pollute the air as a fire burns out of control at a dump*

✳ **Pollution** is the unclean state of the environment (land, atmosphere or water) resulting in physical, chemical and biological changes that can seriously affect ecological systems. Pollution is caused largely by the improper disposal of waste.

Types of pollution

The type of pollution refers to the source from which the pollution originates. In the Caribbean, the primary sources of pollution are from agricultural, industrial and urban activities. Many activities pollute more than one part or element of the environment – land, air or sea.

Agricultural pollution

- Chemicals from fertilisers and pesticides are the most common form of pollution from agriculture. Chemicals find their way into soil and streams, causing the water and soil to become polluted. It was reported in the Jamaica Star newspaper in March 2005 that large quantities of gramazone (a weed killer) were found in a section of the Rio Grande. Water supplies to communities nearby had to be discontinued until tests confirmed that the water was safe for human consumption.

226

- Organic substances from agricultural processing and fertilisers also pollute natural systems. Coffee pulp from processing plants has been a source of pollution in the Blue Mountains. Sugar cane processing produces a waste called dunda, which pollutes waterways and sends a strong odour into the atmosphere. Organic pollutants increase the nutrient load in rivers, causing a sudden increase in plant growth. This reduces oxygen levels in the water, a process known as eutrophication. This process can kill wildlife in rivers (especially fish).

- Clearing of land for cultivation causes erosion and leads to increased sedimentation in rivers and seas (see Chapter 16).

Industrial pollution

- Chemical pollution from industries usually ends up in waterways and along the coast. Many industries in the Caribbean are at or near seaports. Inadequate facilities for disposal mean that waste materials end up in the sea.

- Air pollution from factories and oil refineries is caused by the combustion of the fuels used to supply energy.

- Solid waste and other bulky materials from industries such as scrap metal processing are difficult to dispose of. Sewage and raw effluent from factories sometimes end up in waterways, eventually reaching the sea.

Urban pollution

- Urban populations are the source of most domestic solid waste in the Caribbean. Disposal of this waste is one of the greatest challenges today in the fight against environmental degradation. Large populations (see Chapter 10) and increased consumer demand are producing huge quantities of garbage. Much of this is being dumped in gullies and rivers and eventually along the coast, because official landfill sites are ineffective in handling these huge quantities of waste.

- Improper disposal of sewage is another major source of pollution in the Caribbean. There are not enough sewage treatment plants to deal with the waste produced by the increased populations in cities. Functioning plants are under pressure and so often malfunction, producing strong odours. Some untreated sewage ends up in the waterways.

Areas in the Caribbean where pollution is a problem

Pollution is a serious problem in those areas where agricultural, industrial and urban activities are concentrated. Most industrial activities take place in urban areas. The discharge of solid waste and effluent from industrial and residential sources is therefore a major issue in the cities. Most cities are along the coast, so pollution of coastal waters is a serious problem in many territories. Kingston Harbour in Jamaica is one of the most polluted natural harbours.

- Thousands of litres of untreated sewage flow into Kingston Harbour every day.

- Waste from industrial activities along the harbour front flows into the sea through water channels and pipes leading directly into the sea.

▲ *Figure 15.2 A sewage pipe discharging waste into Kingston Harbour, Jamaica*

- Chemicals from pesticides and fertilisers used on farms seep into rivers like the Rio Cobre and Hope River and eventually end up in the harbour.

- Garbage that has been dumped in gullies is washed into the sea by heavy rains.

➤ *Figure 15.3 Garbage dumped in a gully ends up in Kingston Harbour, Jamaica*

- Each year during the rainy season, tonnes of sediment are washed into the harbour via the Rio Cobre and Hope River. These sediments choke offshore reefs and raise the level of the harbour floor.

There are several other areas in the Caribbean where pollution is a problem, for example:

- Couva Bay in Trinidad: pollution by industrial waste from Point Lisas Industrial Estate and other areas runs off into streams that eventually end up in Couva Bay.

- Caroni Swamp in Trinidad: industrial chemical wastes seep into the swamp.

- Black River in Jamaica: dunda from sugar cane processing ends up in the Black River basin. When the concentration is high, a pungent odour is released into the atmosphere.

- Belize City in Belize: much of the city lies below sea level, so flooding is a major issue at times of heavy rainfall. The canals that take the floodwaters to the sea are also conduits for sewage and waste water. Sewage pollution is therefore a problem along the coastline close to Belize City.

Other areas with pollution problems are discussed later in this chapter.

✳ **Global warming** is a rise in the average global temperature caused by a man-made increase in the level of greenhouse gases.

The natural warming of the Earth

The atmosphere is like a thin blanket of gases that protects the Earth from **solar radiation** (direct heat from sun). When the sun's rays reach the Earth's atmosphere, some of them are absorbed by the clouds and are **reflected** back into space. The remainder reach the Earth's surface and warm it.

➤ *Figure 15.4 How the Earth is warmed*

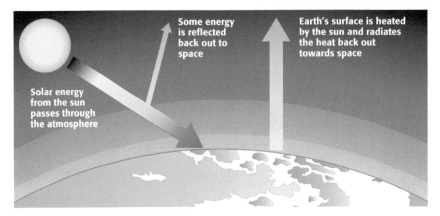

Heat from the Earth's surface rises and warms the air around us (it is this heat that influences weather and climate). Some of the heat that rises escapes back into space as **infrared radiation** and some is trapped by the atmosphere. It is the trapped heat that keeps us warm. This is the natural **greenhouse effect**. The difference between incoming solar radiation and outgoing radiation is known as the **net heat energy**. Any variation in the average amounts of radiation gained or lost can lead to either excess warming or excess cooling, resulting in climatic change and unstable weather conditions.

The greenhouse effect

➤ *Figure 15.5 The greenhouse effect*

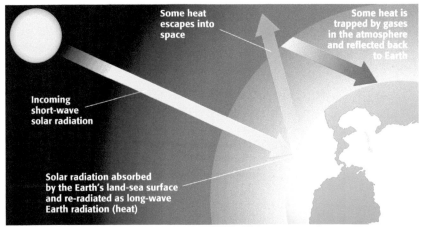

A greenhouse is made of either glass or plastic, which allows sunlight to enter easily. Warm air is trapped inside as the glass or plastic prevents the heat from escaping. The greenhouse heats up, just like a car parked in the sun all day.

Like glass, the atmosphere is relatively transparent to solar radiation and so it allows incoming heat to reach the surface. Some of the gases in the atmosphere, for example carbon dioxide, methane and water vapour, form a blanket around the Earth and trap heat, just like the glass of a greenhouse. The effect is similar to what happens in a greenhouse, so it is referred to as the **greenhouse effect**, and the gases that cause this are called **greenhouse gases**.

If you have ever tried to get into a car that has been parked in the sun all day, on opening the car door you will have felt the heat that has built up inside. The Earth, like the car, stores the heat trapped by greenhouse gases in the atmosphere.

The greenhouse effect is a process by which the Earth is kept warm. However, if the concentration of greenhouse gases increases, more heat is trapped and temperatures rise, causing **global warming**.

➤ *Figure 15.6 Long-term changes in global temperatures*

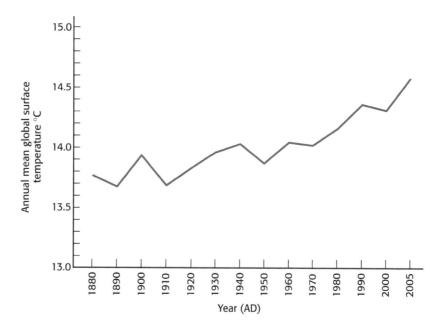

What causes global warming?

The increase in average global temperature is primarily the result of human activities. Both industrial and domestic activities produce and expel excess greenhouse gases into the atmosphere. **Carbon dioxide** (CO_2) is one of the most common greenhouse gases. It comes from the burning of **fossil fuels** – oil, coal and natural gas. Large amounts of CO_2 are emitted daily from factories, motor vehicles, the clearing of forest (deforestation – see Chapter 16) and the production of power from fossil fuels.

➤ *Table 15.1 Greenhouse gases*

Gas	Chemical formula	Some human causes of gas production
Carbon dioxide	CO2	Burning of fossil fuels, solid waste and wood; deforestation resulting in less carbon dioxide being removed from the atmosphere.
Methane	CH4	Decay of organic matter, raising of livestock, extraction of fossil fuels, rice cultivation.
Nitrous oxide	N2O	Use of nitrogen fertilisers, burning of fossil fuels and wood.
Ozone	O3	Air pollution
Halocarbons	HFCs, CFCs, HCFCs	Use in solvents, cleaners and colourants.

We all emit greenhouse gases into the atmosphere when we:

- watch television
- play video games
- open a refrigerator
- microwave a meal
- leave lights on
- use air-conditioning
- charge a cell phone
- use the computer
- drive in a motor vehicle

… and so on. All these activities require energy that uses fossil fuels. Most power stations, especially those in the Caribbean, use gas, oil or coal to generate **electricity**, and CO^2 is released when these are burnt. Vehicular traffic is also growing in Caribbean cities. As people buy more cars and demand better transport facilities, fuel is burnt and more greenhouse gases are emitted into the atmosphere.

If there is too much CO^2 is in the atmosphere then the greenhouse effect will be enhanced, causing temperatures in the atmosphere to rise.

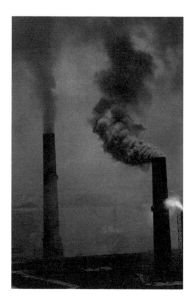

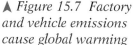
▲ *Figure 15.7 Factory and vehicle emissions cause global warming*

Exercise

1 Conduct a traffic survey of your town or the town closest to you. The survey may be done in half-hour periods and must be done in a group or in pairs. It may be best to do this during peak hour traffic. Prepare a tally sheet and include on it the following:

- types of vehicle (car, truck, motorcycle etc.)
- visibility of fumes from exhausts (dark puffs of smoke)
- vehicles carrying more than one person (car pooling).

After the survey, tally your scores and record your totals.

a Draw a bar graph to show the types of vehicle observed during your survey.

b What was the total number of vehicles tallied?

c How many of these vehicles had visible exhaust fumes?

d Draw a divided circle (pie chart) to show the cars seen with drivers only, and those with passengers.

2 Write a short essay with the title: 'How car pooling could help to reduce traffic and subsequently reduce carbon emissions'.

Population growth and increased **consumption patterns** have increased levels of greenhouse gases. This is because the more people there are, the more **resources** are needed to meet their demands. This creates another problem: for every item used by consumers, waste is produced. The waste going into our **landfill** sites decomposes, creating CO_2 and methane which go into the atmosphere.

Much of this waste is attributed to **modernisation**. We change our cars, furniture, clothing and gadgets much too often. Globalisation has brought a new trend in consumption patterns. There is always something new to purchase. We then discard all the old items for newer designs, even if the old ones still work. The production of **consumer goods** and the disposal of resulting waste materials is now one of the greatest contributors to increased greenhouse gases into the atmosphere. Global warming is now a consumer affair.

Clearance of land for settlement and agriculture, often by burning, has increased greenhouse gases and has also taken away one of the ways in which nature cleans the atmosphere. Plants absorb CO_2 and use it to make their food, but deforestation (see Chapter 16) leads to an excess of CO_2 in the atmosphere.

When we demand meat and dairy products we are also contributing to the greenhouse gases in the atmosphere. Commercial cattle rearing produces methane in the form of animal excreta and from the slaughtering processes.

It is estimated that if human activities continue to increase the level of greenhouse gases, global temperatures will rise during the twenty-first century by 1.4°C to 5.8°C (IPCC, *Third Assessment Report*, WG1).

What is happening in the Caribbean?

The Caribbean islands are small, so human activities may have a greater, faster impact on the environment than they would in some larger, more developed countries. Have you noticed how congested your capital city has become? Next time you are in a car or bus in your capital city, take note of the traffic congestion and the fumes that are produced there.

Another unpleasant change in the Caribbean is the increase in the use of plastic bottles and Styrofoam containers. It is too expensive to recycle the waste from these products, so they end up in city dumps where they are burnt. During burning, harmful gases escape into the atmosphere.

Exercise

1 Conduct an investigation in your neighbourhood to find out the likely **sources of greenhouse gases**. Take a sample of 30 households, or 50 students at your school.

Find out the answers to some of the following questions:

- How many bags of garbage do you generate each week?
- Do you sort and recycle your waste?
- Do you use energy-saving light bulbs?
- How many cars does your household have?
- How often do you eat beef?
- How many televisions are there in your house?
- Do you use Styrofoam containers?
- Would you refuse to use a product if you knew it was harmful to the atmosphere, even though it is safe for you?
- Is your car air-conditioned?
- Is your home air-conditioned?

If your investigation reveals that more than 50 per cent of participants (people in the survey) answer to more than 50 per cent of the questions in a way that indicates a negative effect on the environment, then this could be a subject for your School-Based Assessment research.

2 Conduct a **garbage audit** at your school. Find out how much and what type of waste is generated. Place your results on graphs and post your findings on your school noticeboard. Make suggestions for changes. You can use the 4 Rs of waste management: **reduce**, **reuse**, **recycle** and **refuse**.

Effects of global warming

If global warming continues there will be many consequences:

- The temperatures of oceans will increase.

- Sea levels will rise – Caribbean islands will be very vulnerable.

- Glaciers and ice sheets will melt, for example in Greenland.

- Groundwater near coastlines will become saline – some groundwater sources in Barbados and Jamaica are already affected.

- Coastal flooding will increase, especially during storms – Belize will be seriously affected.

- Beach erosion will increase, affecting tourism in the Caribbean – Negril, in Jamaica is already being affected.

- Some coastal ecosystems, like coral reefs and mangrove swamps, will be destroyed.

- Coral reefs will die as a result of bleaching (see Chapter 16).

- Coastal populations will be displaced.

- The rate at which diseases are transmitted will increase, for example disease-bearing mosquitoes will increase in number.

- Economies will be weakened, especially if tourism and agriculture are affected.

- More and bigger storms will develop because of the warmer seas.

- Weather patterns will become irregular and extreme, for example long droughts or very heavy rainfall (see Chapter 14).

- Many forests will be lost to desertification.

➤ *Figure 15.8 Beach erosion due to a change in sea level*

It is important to remember that the Caribbean is made up of predominantly **small island states**. This makes them especially vulnerable to the impacts of global warming. Even though many islands are mountainous, most of their towns and villages are along the coastline. If global warming continues to increase, some Caribbean islands like the Bahamas and Barbados could disappear, or at least have most of their towns wiped out by the sea.

Consequences of global warming on Mauritius

Mauritius is a small volcanic island in the Indian Ocean about 800km east of Madagascar. With a size of 1268.2 sq. km (788 sq. miles) and a population of 1.23 million, it has one of the highest population densities in the world (600 people per sq. km). As a small island developing state (SID), Mauritius is among the most environmentally vulnerable countries of the world. It is geographically remote (alone in the ocean), with little topographical height (highest point is 828m above sea level) and has limited natural resources. It is therefore exposed to the impacts of global warming and sea level-rise in a similar way that Caribbean islands are.

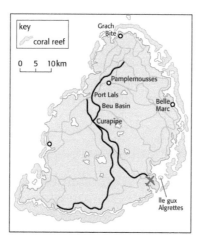

➤ *Figure 15.9 A map of Mauritius*

While scientists are still gathering data on the incidence and impacts of global warming and climate change, Mauritius is already experiencing the consequences.

- The Indian Ocean is experiencing an increase in tropical cyclones – to about 10 per year.

- Evidence of sea-level rise is being observed in the groundwater, where saline intrusion is becoming common; there is a shortage of fresh water in Mauritius as a result.

- Changes in tidal height could make some coastal lands uninhabitable – populations near the coast will be the first to be affected.

- Mauritius could face extinction in this century if rates of sea-level rise increase.

- Storm surges will increase wave action – beach erosion is already evident along many beaches in Mauritius.

- Mauritian coral reefs are already degrading.

- Abnormal tidal ranges are occurring.

Mauritius taking action

Recognising its vulnerability, Mauritius has developed organised management schemes to deal with environmental issues.

- It is party to various international environmental agreements such as The Kyoto Protocol on climate change, Marine Life Conservation, Ozone Layer Protection, Ship Pollution, Law of the Sea, among many others.

- It has developed several organised management schemes, one of which is a waste management strategy to facilitate the handling of materials such as glass, plastics, metals, etc.

- The International Meeting for the 10-year Review of the Barbados Programme of Action for the Sustainable Development of Small Island States was held in Mauritius (SIDS – Mauritius 2005). This facilitated talks and potential for funding to aid Mauritius in the many environmental problems listed above.

For more information on Climate Change in SIDS see
http://www.un.org/smallislands2005/

▼ *Table 15.2 Causes and effects of global warming*

Gases	Sources	Impact on climate
Carbon dioxide	• Burning of fossil fuels (oil, coal, natural gas) • Flaring of natural gas • Changes in land use – deforestation for agriculture • Manufacture of cement	• Intensifies heat-trapping properties of atmosphere • Changes rainfall pattern • Increases regional and global temperatures
Carbon monoxide	• Motor vehicle emissions • Factories (e.g. aluminum plants)	• Increases local/regional/global temperatures
Sulphur dioxide	• Coal-burning power stations, heating of buildings • Cooking gases • Factories • Transport	• Contributes to acid rain
Methane	• Garbage in landfill sites • Wet-rice farming • Coal mining • Sewage plants • Cattle farming • Leaks from natural gas lines • Biomass	• Methane traps about 25 times more heat than CO^2 and could become the most significant greenhouse gas
CFCs, HCFCs, PFCs, SF6	• Refrigerators and car air-conditioning • Aerosol propellants • Insulation	• Degrades and reduces ozone layer in upper atmosphere • Thinning of upper ozone layer may cause cooling in lower atmosphere • Thinning also allows penetration of more ultraviolet rays
Nitrogen Oxide	• Coal-burning power stations • Combustion from factories and (especially) high-performance motor cars • Fertilisers	• Wastes react with volatile organic compounds in the presence of sunlight and heat to form smog • Causes acid rain

Write an essay comparing the similarities in consequences of global warming on the Caribbean with that of Mauritius.

➤ *Figure 15.10 Sources of greenhouse gases in the atmosphere*

Country	Total greenhouse gas emissions in 2000 (million tonnes CO2 equivalent)
Australia	491
Canada	675
China	4942
Eritrea	1
France	512
Guyana	4
Jamaica	13
Japan	1333
Mexico	511
Sudan	96
Trinidad & Tobago	22
United Kingdom	660
USA	6924

Use Figure 15.11 to answer the following questions.

1 Which country is the largest producer of greenhouse gases?

2 Which country produces the least?

3 Suggest reasons for the distribution shown.

Measures to reduce the impact of global warming

We need to do the following:

- raise environmental awareness through education

- create laws to protect the environment, e.g. impose stricter penalties for removing trees, or reward efforts to replant

- recycle and reduce waste

- find alternative sources of energy, e.g. renewable sources like hydro, solar and tidal power

- raise consumer awareness, encouraging people to refuse to buy some products

- improve domestic use of energy, e.g. conserving electricity

- improve public transport so as to reduce the use of private vehicles.

Some **more developed countries (MDCs)** have greater resources and political power to correct and reduce the impact of environmental problems. In 1987, a group of countries gathered in Montreal, Canada, to listen to reports on the dangerous effects of greenhouse gases on global climates. A decision was made to reduce the use of CFCs (chlorofluorocarbons, a pollutant found in some aerosols and refrigeration units) as a way of reducing the impact of global warming. This agreement is called the **Montreal Protocol**. In 1990 the agreement was revised in London, from a reduction technique to a total phase-out process. Many countries agreed not to produce or use any items that produce CFCs by the year 2000. Jamaica was one of the **signatories** to the agreement.

However, many countries in the developing world cannot afford to take such measures against global warming and so are not able to sign up to such agreements.

Reducing the impact of global warming in Canada

Alternative energy sources are costly to set up. Some areas need much research and testing. For example, Canada, a more developed country (MDC), has both the resources and technology to explore and develop alternative sources of energy. Much of Canada's industrial energy comes from hydroelectric plants. This reduces the amount of greenhouse gases that would otherwise be emitted into the atmosphere.

Canada has put in place measures such as:

- use of renewable energy technology – water, wind and tidal

- conservation of forests reserves – large land areas left unspoiled

- reforestation programmes

- municipal recycling

- a developed transport sector

- environment laws and enforcement programmes
  - Canadian Environment Protection Act
  - vehicle emissions laws
  - environmental advocacy and lobby

- conventions and protocols (international agreements)
  - Canada–US Agreement on the Trans-boundary Movement of Hazardous Waste
  - The Montreal Protocol on Substances that Deplete the Ozone Layer

- environmental law resources – The NAFTA Commission for Environment is based in Canada.

These websites may be explored to give further information on Canada's position on greenhouse gas emissions and control:
www.ec.gc.ca/ele-ale/know/know_e.asp
www.ec.gc.ca/envlaw_e.html
www.brunel.ac.uk/research/cer/resource/law.htm

How does the average Canadian citizen take part in all this?

- Garbage is recycled.

- Public transport is used by everyone.

- Car pooling is encouraged.

- Car up-keep is expensive (insurance, registration, maintenance and gas), which deters young people.

- People obey the environmental laws.

- Environmental education is part of the school curriculum.

- 'Green' organisations have power and can influence the government (a **green** organisation is a group that supports environmental protection).

Environmental degradation

By the end of the chapter students should be able to:

- explain the causes and effects of **coral reef destruction**
- describe **measures** taken to reduce coral reef degradation **in one Caribbean territory**
- define **deforestation**
- explain the **causes** and harmful **effects** of deforestation
- explain **measures** used to reduce **the impact of deforestation in one Caribbean territory**.

The Caribbean was one of the most beautiful places on earth when the early European explorers arrived here. Thick forest, cascading waterfalls and pristine beaches were typical sights on the islands of the West Indies. However, as the population grew, human actvity began to affect not only the land but the coastal environment too.

Coral reef degradation

For this section it will be important to link to the information in Chapter 7 on coral reef formation.

Causes and effects of coral reef degradation

You may not have realised it but forests are closely connected to coral reefs. What we do on the land affects what goes into the sea. The *'Ridge to Reef'* concept, which will be discussed later on in this chapter, emphasises the reality of this.

Virtually all coral reefs in the Caribbean have been affected by humans. Some are in such a deplorable condition that the effects are having a far-reaching impact on local ecology and the economic viability of tourism. Most of these fringing reefs lie very close to shore and so are badly affected by the dense human settlement and economic activities that exist along Caribbean coastlines (See Chapter 10). There is more human activity along the coast than in the interior

▲ *Figure 16.1 The distribution of coral reefs in the Caribbean region*

of the islands, and where activities exist in the mountainous interior, these seem to be unsustainable. Some of the recognisable activities by humans are described on the following pages.

▲ *Figure 16.2 A damaged coral reef covered in algae*

▲ *Figure 16.3 A damaged coral reef covered in silt*

▲ *Figure 16.4 A damaged coral reef suffering from bleaching*

- **Pollution from sewage discharge:** sewage pollution is one of the major causes of coral reef degradation. Sewage is rich in nutrients that foster the growth of excess algae, phytoplankton and bacteria from faecal coliform. Where sewage treatment plants do exist, they often function poorly, and so the problem recurs.

Exercise

Find out if the communities in which you live have a functioning sewage treatment plant.

- **Industrial and agricultural run-off:** toxic chemicals and organic waste from industrial and agricultural activities have found their way into Caribbean waters from both direct and indirect run-offs. In Trinidad, coastal pollution has become a serious problem, especially in the Gulf of Paria. Many of the manufacturing plants along Trinidad's western coast discharge waste into the Gulf. In the manufacturing of petrochemical products, ammonia and phenols are widely used and subsequently discharged into the sea. These chemicals are toxic to marine life on the coral reefs. They also change the pH balance of the water.

Offshore drilling of oil poses a threat to the growth of corals off the coast of Trinidad. Run-off from industrial and agricultural activity in Venezuela is also a threat to these reefs.

➤ *Figure 16.5 Industrial areas along the west coast of Trinidad*

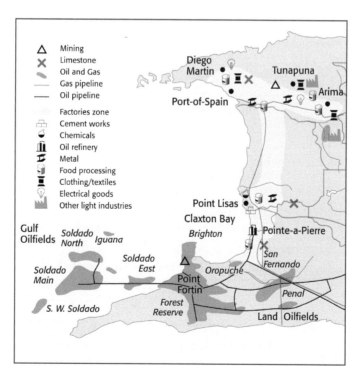

Considering the extent of agricultural activities in the Caribbean, you can well imagine that the use of pesticides and fertilisers is prevalent. These reach the sea from either direct run-off or illegal dumping. Run-off from farms in the interiors enters river systems that eventually reach the sea. Organic fertilisers increase the growth of harmful algae.

- **Tourism-related activity (coastal development):** tourism has contributed to major development in the Caribbean. However, tourism activities have contributed to the degradation of many coral reefs in the Caribbean. Destruction from tourism has taken place in many forms:

  - *Clearing of land for construction of hotels, roads and artificial beaches* – the removal of trees has led to erosion and subsequent sedimentation of sea water.
  - *Artificial beaches* – the introduced sand is easily removed by wave action and disturbs the clarity of the water.
  - *Improper sewage disposal or insufficient sewage treatment infrastructure for resort areas* – in Negril, the algal growth has been blamed for the reduction of live reef cover from up to 80% to 5%
  - *Collection of specimens for souvenirs*
  - *Berthing of boats* – boats anchor on reefs to allow scuba diving and viewing.
  - *Trampling by divers.*

▲ *Figure 16.6 Hotel under construction*

The Negril, Montego Bay and Ocho Rios areas together make up the largest tourist destinations in Jamaica. The coral reefs in these areas are all damaged and have now become a costly concern. The unusual and rapid erosion of the seven-mile stretch of beach in Negril is being blamed on the damage to the coral reef.

- **Inland pollution and erosion (siltation) from poor land-use practices:** sediment from land has a negative effect on corals. When rain falls, most of the water flows along the surface to the sea – the '*Ridge to Reef*' concept. Soil that is eroded from hillsides finds its way to coastal waters and chokes corals. Have you ever noticed a brown colour around the coastline after it rains? Well, that is silt and mud. The more this happens the more affected the corals will be as silt chokes the corals and prevents sunlight from entering. All poor land-use practices that remove trees without proper mitigation influence soil erosion and surface run-off.

  Rubbish dumped in gullies and rivers is washed into the sea during heavy rainfall. Layers of debris can be seen covering the sea after and during a storm. Some of these get stuck in the coral reef after the tide or surge subsides. They do irreparable damage to coral reefs.

- **Over-fishing:** the use of irregular-sized nets, poison and dynamiting are some of the more harmful practices that damage coral reefs. Over-fishing can lead to the loss of very important fish species. This subsequently disturbs the whole ecological balance and again leads to destructive algal bloom.

- **Natural causes of coral reef degradation:** coral reefs may also be destroyed by natural occurrences such as hurricanes. The strong waves associated with hurricanes and storms erode the most fragile section of the reef – the top. This however can grow back naturally. High water temperatures 'bleach' corals (make them white), eventually killing them. Sometimes natural diseases will wipe out or reduce species that are vital to the functioning of the reef, for example, sea urchins.

- **The implications of coral reef degradation:** coral reefs protect the land from storm and tidal surges and provide a source for economic development and fish for consumption. If they are damaged then the implications could seriously affect sustainable development in the Caribbean. So it is very important that we protect them.

▼ *Figure 16.7 Degradation of the land increases surface run-off*

More information about the state of Caribbean reefs can be gained from World Environment News (WE) website (http://www.thewe.cc/content/more/archive/july2003/80_percent_of_coral_caribbeanreef_destroyed.htm).

▲ *Figure 16.8 Degradation of coral reefs may result in increased coastal flooding*

How can we protect our coral reefs?

To protect the coral reefs we have to desist from, or reduce, the activities that will destroy them. Alternatively, safe environmental practices in the activities that were previously listed must be adhered to. Caribbean countries, collectively and individually, must take an approach of sustainable development towards protecting coral reefs.

Measures to reduce coral reef degradation – the case of Jamaica
Coral reefs are one of Jamaica's important resources. The north and east coast is fringed with a narrow shelf of diverse reef systems. Various studies have shown that the reef systems around the island have been rapidly deteriorating, primarily because of human-induced processes. Here is a list of some **measures** that are being used to reduce coral reef degradation in Jamaica. Many of these arose from the establishment of the Natural Resources Conservation Authority (NRCA) in 1991 and the environmental summit in Brazil, popularly called the Rio Earth Summit, in 1992.

- **Signature to international protocol:** in 1994 a number of nations came together to launch the International Coral Reef Initiative (ICRI) to protect coral reefs in partnership with coral reef nations. Jamaica is a signatory to the ICRI. The University of the West Indies supervises Jamaica's monitoring programme.

- **Policies and legislative framework:** these have been put in place to monitor and prevent further degradation of coral reefs in Jamaica:
  - Natural Resources Conservation Authority Act (1991)
  - Policy for Jamaica's System of Protected Areas (1997)
  - Towards a National Strategy on Biological Diversity in Jamaica (2001)
  - Integrated Watershed and Coastal Zone Management Branch (2004) – developed out of the '*Ridge to Reef*' concept where the Watershed Unit and Coastal Zone Management Branch in NEPA merged.

  This legislative framework provides a means for enforcement so that if a person or company is found in breach of any of the above, they can be charged under the law.

- **Marine Parks or Marine Protected Areas**: Jamaica has established marine park areas to help with the management of coastal zones.
  - Montego Bay Marine Park
  - Negril Marine Park
  - Ocho Rios Marine Park
  - Palisadoes/Port Royal Marine Park.

  Under the Policy for Jamaica's System of Protected Areas, there is restriction on development in the above-mentioned areas. Where development is necessary, a licence must be sought.

- **Awareness and education campaigns**: in Jamaica, non-governmental organisations (NGOs) such as the Negril Area Environmental Protection Trust (NEPT) and Friends of the Sea (FOTS) have embarked on area-specific environmental education programmes for schools and the general public aimed at specifically protecting reefs along the Negril and Ocho Rios coastlines.

- **Management programmes**: several projects and programmes have been put in place by the National Environment and Planning Agency (NEPA) and private environmental non-governmental organisations (NGOs) to manage local coastal zones:
  - Negril Area Environmental Protection Trust (NGO)
  - Montego Bay Marine Park (NGO)
  - Friends of the Sea (NGO)
  - Negril Coral Reef Preservation Society (NGO)
  - Fisheries Division (governmental).

  NGOs manage these areas through the use of rangers and enforcement officers as well as integrated education programmes. Some of these manage the parks with funds from donor agencies that require strict reporting of outcomes to the programme. This helps to ensure the success of the programmes.

- **Environmentally safe practices**: some hotels have now been certified by global environmental organisations for the conservation methods they employ in the running of the resort. The Negril Tree House and Negril Cabins are Green Globe certified.

➤ *Figure 16.9 Montego Bay Marine Park is a protected area: take note of the fringing reef*

© Montego Bay Marine Park

● **Alternative tourism:** heritage and ecotourism are being encouraged as part of a sustainable tourism plan. The Blue and John Crow Mountains National Park (BJCMNP) has an interpretive hiking programme that provides guided tours throughout the park and to the peak. The Cunha Cunha Pass is a trail that cuts through the park in the mountains with a rich Maroon heritage. This is an alternative to the sand, sun and beach concept of tourism.

Exercise

Work in groups for this exercise. You are the directors of a local NGO and have been given management responsibilities for a stretch of coastline in your area:

1 Write a list of the problems you may need to solve.
2 Discuss four ways you can go about raising awareness about protecting coral reefs.
3 Make a poster to publicise your message. The poster should include
   a the name of your organisation,
   b the major problem affecting the coastline you're managing
   c the message you want to give out.

Deforestation

✳ **Deforestation** is the cutting down of trees and clearing of natural forest.

➤ *Figure 16.10
Deforestation on the lower slopes of the Blue Mountains, Jamaica*

Exercise

Do this exercise now, then review and evaluate your findings once you have completed the rest of the chapter.

The concept of sustainable development is debatable. Your teacher will divide you into groups for a debating competition. The motion you should debate is: *This house believes that environmental degradation is just an exaggeration – some resources must be expended so that humans can exist.*

Define the terms **sustainable development** and **environmental degradation** in your discussion.

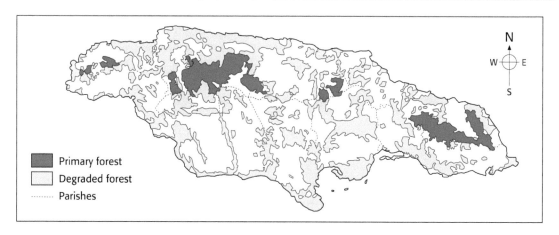

Primary forest
Degraded forest
········· Parishes

▲ *Figure 16.11 Forest cover in Jamaica – present*

Causes of deforestation

Deforestation takes place daily on large and small scales in the Caribbean. It is primarily carried out because of rapid population growth that results in increased human activities.

- **Agriculture:** one of the most common and localised causes of deforestation in the Caribbean is the need for land for arable and pastoral farming.
  - *Small farmers* using **shifting cultivation** clear entire hillsides as they move from place to place in search of fertile land. The entire Buccament valley in St Vincent was deforested by banana farmers and is now being restored by the St Vincent and the Grenadines (SVG) Forestry Department.
  - *Coffee farmers* in territories such as Jamaica seek hilly slopes that have a cool climate to cultivate coffee. However, the hilly slopes need trees to maintain soil stability.

- *Yam cultivation* requires the use of sticks to support the plants. In Jamaica, a phenomenal amount of tree saplings (young trees) are removed annually from the interior of the Cockpit Country for use as yam sticks. The Southern Trelawny Environmental Agency (STEA) reported that an average of 6–9 million tons of tree saplings are removed annually from the parish of Trelawny (http://www.stea.net).

- **Charcoal and fuel wood production:** many Caribbean communities still rely on wood and charcoal for domestic fuel. Trees are removed indiscriminately from some of the most fragile forest environments. Rural communities on the fringes of forests take advantage of the forest as a resource.

- **Industrial and commercial development:** second only to agriculture is the need for land for development of the built environment. Developments now exist where tropical and mangrove forests have been cleared for land reclamation. The large Point Lisas Iron and Steel Industrial Estate on the west coast of Trinidad and the Marcus Garvey industrial area in Kingston, Jamaica are built along reclaimed wetlands.

- **Residential development:** new housing developments are on the rise as population growth increases in the Caribbean. Examples of new settlements include Portmore in Jamaica and Westmoorings in Trinidad. These areas are in recently forested wetlands. Look back to Chapter 11 and see if you can make the link.

- **Tourism:** the activities of tourism demand large acreage of land for hospitality, sports and entertainment. Mangrove forests are cleared for beachfront properties and wetlands are filled in for added land space. This is the situation in west and south Barbados, St Lucia and Jamaica. In almost the entire Caribbean the narrow strips of coastal plains are fully developed into tourism resorts.

- **Road and highway construction:** the growth of cities and towns along the coasts and hinterlands of the Caribbean creates the need for more roads, as people and goods need to be connected and move from place to place. New highways have been constructed in Barbados, Jamaica and St Lucia over the past decade, accounting for thousands of acres of deforestation.

- **Illegal farming:** forests have fallen prey to a new wave of illegal land tenure. The cultivation of marijuana has accounted for large acres of land being deforested as marijuana farmers seek to carry out their activities in remote environments. The forestry departments of Jamaica and St Vincent have been grappling with this problem.

▲ ➤ *Figure 16.12 Major causes of deforestation in the Caribbean*

- **Timber extraction:** timber extraction in the Caribbean is limited given the nature of species diversity, denseness of forest and, in many cases, inaccessability. The majority of timber is from hardwood varieties. However, new species of softwood varieties have been introduced.

From the list above (which is by no means exhaustive) it is clear that human activities rely on, but at the same time destroy, the forest. The forest is very important and if not properly managed can be easily damaged, with far-reaching effects on the wider environment. From Chapter 9, you should be able to make a list of all the important functions of trees in the environment.

Exercise

The following is an experiment that must be worked out with your teacher:

1 Count the number of fully grown trees in your school.
2 Find out their names and list them.
3 Place them into groups of endemic and introduced species.
4 Collect and draw samples of leaves and seeds.
5 Try to grow small seedlings in a plastic or glass container (reuse an old one).
6 Transplant one seedling to a pot and keep it in the classroom.
7 Transplant the rest of the seedlings around your school compound.
8 Record the date in your notebooks and then watch the seedlings grow – you will have left something valuable from your geography lesson.

Effects of deforestation

The destruction of forest in the Caribbean has had several effects on individual territories and on the region as a whole.

➤ *Figure 16.13 An eroded hillside*

- **Soil erosion:** the thinnest uppermost layer of the earth is called the topsoil. It is like the 'skin' of the earth, very thin and fragile. It can be protected only by trees. Millions of tons of topsoil are lost annually in the Caribbean as a result of deforestation. Soil erosion leads to further problems such as:

  - *Infertile soil* – topsoil holds the humus (nutrient-rich organic matter) that is most valuable to the fertility of the soil. When this is washed away, the infertile underlayer is left exposed. In St Vincent and the Grenadines infertile soil is visible on bare slopes in the Cumberland region. The colour and cohesiveness of the soil changes when vegetation cover is removed. This effect is constant around the Caribbean where hill slopes are cleared of vegetation for farming.

  - *Siltation of riverbeds and streams* – large amounts of sediment can be seen after moderate to heavy rainfall in many rivers. The heavy silting in the Yallahs River in Jamaica after each rainfall incident is a result of deforestation in the hills above. (See Chapter 6.)

  - *Damage to coral reef* – coral reefs need clear waters, as you learnt in Chapter 7. Soil that is washed into the sea makes the water turbid, that is, cloudy (this is explained earlier on in this chapter under *Coral reef degradation*).

- **Rapid water run-off:** trees absorb excess water through their root systems, reducing the amount of water that stays on the surface. Therefore, water will move much faster and in larger volumes down bare slopes. This creates several problems on and below the hill:

  - *Reduction of watershed* – watersheds are in the most forested areas on ridges and hill slopes. These areas become vulnerable to deforestation as humans penetrate forests for land and wood.

  - *Uneven flow of springs and streams* – if the watersheds are damaged then the supply of water to springs and streams will decrease. Many perennial rivers have become temporary streams as a result of deforestation.

  - *Reduction of groundwater supply* – as most of the water runs off and less is percolated into the ground, the water table falls and some groundwater supplies have even disappeared.

  - *Increased chance of flooding* – rapid water run-off leads to a greater chance of flooding in low-lying areas. Areas that were already prone to flooding receive higher than normal levels of water after heavy rains. In September 2004, an entire region of Haiti was flooded as high volumes of water and mud from Tropical Storm Jeanne rolled off the bare hillsides and on to the plains covering the city of Gonaives. Hundreds of people died as a result of these floods. Haiti, which was once covered in forest, has only about 3% of its tree cover left.

➤ *Figure 16.14 A destroyed mangrove forest. Deforestation was a huge factor in the disaster at Gonaives, Haiti in September 2004*

- **Decrease in biodiversity and loss of endemic species:** many animals and flowering plants rely on the forest for survival. Tropical forests, especially, have greater biodiversity than other types of forest ecosystems. When they are destroyed, humans suffer in the long term:
  - *Rare and endemic species are lost forever*, some of which provided food and medicine for humans. Many were a valuable part of the food chain and therefore their reduction means a threat to other species or increase in population of undesirable species.
  - *Reduction of wildlife habitat and population* – hundreds of wildlife species lived in Caribbean forests. The Caribbean, for example, is renowned for the great diversity in bird species; many of which are now endangered or on the verge of extinction. The West Indian Whistling Duck is on the global endangered list because of the destruction of marine wetlands.

  ❋ **Biodiversity** is the variety of species of plants and animals that exists in an area and forms the ecosystem of that area.

- **Coastal damage by removal of mangrove buffer strip:** one of the primary functions of mangrove forest is to protect the coastline from storm and tidal surges and protect the sea from surface run-off (siltation). Many coastlines are being eroded because of the absence of this buffer zone, and storms are having a greater effect on coastal settlements.

- **Loss of natural beauty:** the diversity of flora and fauna found in the Caribbean has been attracting visitors for years. However, as development removes these, the beauty of the West Indies is disappearing. The tropical paradise is now being replaced with concrete, dry rivers and murky seas.

The effects listed above are all interrelated. No one effect occurs independently of the other, instead one effect triggers the other.

Exercise

Make a poster illustrating one of the following themes:
- Types of bad land use that destroy forest environments
- Why we must love our trees
- Individual action matters
- Save our trees

The best three could be displayed on your classroom noticeboard.

▲ *Figure 16.15 The impact of bad land use (deforestation) on the land*

Interdependence between human activity and the environment creates conflict; so the two have to come together in a sustainable way for the survival of both. Unless we pay attention to the environment there will be nothing to gain resources from: we will have 'killed the goose that lays the golden egg'. The very environment we destroy to create revenue is the environment we need to sustain life and long-term income.

▼ *Figure 16.16 Forest evolution in St Vincent and the Grenadines before and after deforestation*

▲ *Figure 16.17 Forest evolution in St Vincent and the Grenadines after reforestation*

What is being done to reduce deforestation in the Caribbean?

The case of St Vincent and the Grenadines (SVG)

In an attempt to regenerate forests in the Caribbean, many territories have implemented comprehensive forest management programmes. The Forestry Department in St Vincent and the Grenadines (SVG) has embarked on a massive forest rehabilitation programme which includes the following:

- **Fire Resource Conservation Act passed in 1997:** this legislation protects areas zoned as forest reserves. Anyone found in breach of this Act is liable before the courts.

- **Reforestation of three major watersheds previously damaged:** the programme has embarked on replanting indigenous species such as penny piece, wild plum and board wood in Cumberland, Buccament and Colonarie, three of the damaged watersheds. These areas are monitored by forest rangers who have been trained especially for the rehabilitation programme. Constant evaluation is undertaken to monitor the improved environment and to reduce soil erosion in other areas.

- **Implementation of a forest education programme:** posters, pamphlets and flyers have been created to educate the public in a national drive to restore forest environments through the reforestation of private and public lands. Extension officers have been trained to educate and work with farmers in applying sustainable farming practices such as:

  - Terracing   ● Agro-forestry   ● Inter-cropping.

  Environment education programmes are now being implemented into the primary school curriculum to include forest education. NGO groups such as *Avian Eyes* have been working with schools and training teachers in forest education.

- **Ten thousand acres of land have been set aside for park conservation and saving the parrot:** the parrot species of SVG is endangered, as are most parrots in the region. The park conservation area is set up specifically to encourage the population of parrots and other endangered birds. This will also be used to develop the concept of ecotourism.

- **Ecotourism:** this is the development of sustainable tourism through the promotion of tours, hikes and bird-watching and preservation of SVG's heritage. The UNESCO-funded Youth PATH (Poverty Alleviation through Tourism and Heritage) programme promotes the use of indigenous culture and environmentally sustainable projects. Youth PATH organises cross-island walks and guided forest tours.

The Iwokrama Project in Guyana

The Iwokrama International Centre for Rain Forest Conservation and Development is an autonomous non-profit institution established by the government of Guyana and the Commonwealth. The Centre manages the nearly one million acre (371 000 hectare) Iwokrama Forest in central Guyana. Its aim is to show how tropical forests can be conserved and used sustainably to provide ecological, social and economic benefits to local, national and international communities (http://www.iwokrama.org).

The Iwokrama Forest is of national and international significance. It is an area of high biodiversity in both plants and animals. The government of Guyana has established a large section of this forest as a Wilderness Preserve (protected area). This means that there are legislated policies that restrict use and development of the area in order to protect the forest.

Other Caribbean countries have taken initiatives similar to St Vincent and the Grenadines and Guyana. Case studies are available online. There is generally plenty of information on the internet about deforestation in the Caribbean – find out if your local forestry department has a websites. You could start looking for information on the following sites:

http://www.jcdt.org   http://www.stea.net   http://www.iwokrama.org
http://www.forestry.gov.jm   http://www.caribzones.com/crep.html

Exam practice question

1 a Describe two specific causes of deforestation in a named
            Caribbean country.   *4 marks*
       b Locate and name an example of each cause in the territory
            named above.   *2 marks*

2 Name two ways in which coral reefs are negatively affected by
       human activity.   *2 marks*

3 For a named Caribbean country, explain carefully three
       measures taken by the country to conserve its coral reef
       or to lessen coral reef degradation.   *6 marks*

Exam practice questions

Human–Environment systems

Paper 1: Multiple choice

1 Natural hazards can have devastating impacts on human society. Which of the following are some common impacts in the Caribbean?
 (i) Destabilisation of economies
 (ii) Displacement of communities
 (iii) Disruption in the agricultural sector
 (iv) Destruction of the infrastructure
 A (i) and (ii)
 B (ii) and (iii)
 C (iii) and (iv)
 D All of the above

2 In the long term, volcanic eruptions can have beneficial effects to a country by:
 (i) relocating people to new areas in the country
 (ii) producing deposits of precious metals and mineral deposits
 (iii) providing alternative energy sources
 (iv) bringing new materials to the Earth's surface which produce rich, fertile soils.
 A (i) only
 B (i), (ii) and (iii)
 C (ii), (iii) and (iv)
 D All of the above

3 Which of the following is the most common cause of natural disaster in the Caribbean?
 A Volcanic eruptions
 B Hurricanes
 C Earthquakes
 D Tsunamis

4 Which of the following is the result of secondary effects of an earthquake?
 A Fire
 B Thunderstorms
 C Cracks in buildings
 D Landslides

5   In advance of an approaching hurricane, endangerment to lives can be reduced mainly by:

(i)   relocating people to shelters

(ii) evacuating people living in low-lying areas

(iii) encouraging people to stock up on supplies

(iv) encouraging people to purchase stand-by generators.

A (i) and (ii)

B (iii) and (iv)

C (i), (ii) and (iii)

D All of the above

6   Which of the following has caused **rapid** decline in healthy coral reefs in the Caribbean?

A Sewage discharge in coastal waters

B Global warming

C Overfishing

D Snorkelling

7   Agriculture can contribute to coral reef degradation through:

(i)   filtration and seepage of pesticides and fertilisers

(ii) soil erosion caused by slash and burn methods

(iii) overcropping

(iv) monoculture.

A (i) only

B (i) and (ii)

C (iii) and (iv)

D (iv) only

8   Which of the following are measures best used to protect coral reefs?

(i)   Encourage snorkelling to develop appreciation

(ii) Educate about coral reefs

(iii) Build artificial reefs

(iv) Destroy wetlands

A (i) and (ii)

B (ii) and (iii)

C (i) and (iv)

D (iii) and (iv)

9   Which of the following is a common solid waste problem in the Caribbean?

A Dumping in gullies and along coastlines

B Illegal hospital dump sites

C Pungent odours from dump sites

D Lack of receptacles for garbage

10 Which of the following is the most common greenhouse gas?
   A Methane
   B Sulphur dioxide
   C Carbon dioxide
   D Ozone

11 In the Caribbean, global warming is most likely to cause:
   A an increase in the use of boats
   B an increase in population
   C an increase in air conditioning
   D a rise in sea level.

12 Air pollution in the Caribbean is primarily consequence of:
   (i) vehicle emissions
   (ii) industrial activity
   (iii) radioactive emissions
   (iv) landfills and dump sites
   A (i), (ii) and (iii)
   B (ii), (iii) and (iv)
   C (i), (ii) and (iv)
   D All of the above

13 Which Caribbean country has the least problem with deforestation?
   A The Bahamas
   B Jamaica
   C Haiti
   D St Vincent and the Grenadines

14 The main cause of deforestation on mountainous slopes in the Caribbean is:
   A industrial development
   B residential development
   C agriculture
   D mining.

15 One of the primary reasons that Caribbean countries engage in reforestation programmes is to:
   A provide employment
   B ensure soil and forest conservation
   C increase agro-forestry
   D improve agricultural practices.

Total 15 marks

Paper 2

1 Study the table below, which shows the sources of pollutants reaching the oceans each year, then answer the following questions.

➤ *Table P.1 Sources of pollutants entering the sea*

Source	Percentage
Industrial waste	36
Offshore drilling	2
Domestic waste	45
Seeps and run-offs from farms	8
Construction	9

(a) (i) Name one type of diagram that may be used to show the information given in the table.

   (ii) Which is the main source of pollutants entering the sea?

   (iii) Which source contributes in the least way to pollution of the sea? (3 marks)

(b) If the pollutants entering the sea amount to 3 200 000 metric tonnes, how many metric tonnes came from industrial wastes?

    (2 marks)

(c) (i) Define global warming. (2 marks)

   (ii) List THREE ways in which global warming may have an impact on the Caribbean. (3 marks)

   (iii) Describe THREE measures taken by a more developed country you have studied to reduce the impact of global warming. (6 marks)

   (iv) Describe TWO ways in which deforestation has contributed to environmental damage in a named Caribbean country. Then explain TWO ways in which the named country is addressing this issue. (8 marks)

Total 24 marks

2   Study the diagram below, which shows the hazards caused by hurricanes, then answer the following questions.

Hazzards caused by Hurricanes

➤ *Figure P.1 Hazards caused by hurricanes*

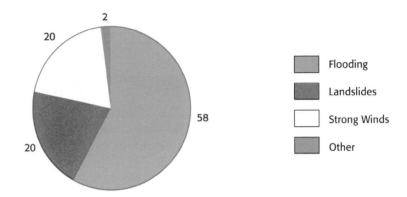

(a) (i)   What is the greatest hazard caused by hurricanes, and by what percentage?

(ii)   What percentage of hazards do landslides contribute?

(iii) Which is the third largest hazard caused by hurricanes?
(4 marks)

(b) (i)   Describe two harmful effects of natural disasters on a named Caribbean country.   (6 marks)

(ii)   Explain one measure taken to reduce the effects described in (i).   (3 marks)

(c)   Name one Caribbean country that has been evacuated as a result of major volcanic eruptions.   (1 mark)

(d) (i)   Describe THREE necessary precautions that must be taken before a hurricane strikes, in order to reduce the effects it may have on the land that it passes over.   (6 marks)

(ii)   Explain TWO measures that must be taken after the passage of a hurricane.   (4 marks)

Total 24 marks

3 (a)

   (i)   Draw a map of a Caribbean territory to show where pollution
         is a major problem.   (4 marks)

   (ii) Describe TWO ways deforestation and urbanisation may lead
         to flooding.   (4 marks)

   (iii) Describe TWO ways in which tourism can lead to the
         destruction of coral reefs.   (4 marks)

(b) (i) Describe 2 ways in which deforestation may lead to a rapid
         loss of soil fertility in equatorial regions.   (8 marks)

   (ii) Name one area of extended coral reef in the Caribbean.

                                                                (1 mark)

Total 24 marks

(c) Explain ONE way in which coral reefs are important to the
    development of beaches in the Caribbean.   (3 marks)

Index